Power PRACTICE™

Timed Tests

Addition and Subtraction

Editor: Carla Hamaguchi
Designer/Production: Moonhee Pak/Rosa Gandara
Cover Designer: Barbara Peterson
Art Director: Tom Cochrane
Project Director: Carolea Williams

Table of Contents

Introduction

Each book in the *Power Practice*™ series contains dozens of ready-to-use activity pages to provide students with skill practice. The fun activities can be used to supplement and enhance what you are already teaching in your classroom. Give an activity page to students as independent class work, or send the pages home as homework to reinforce skills taught in class.

The pages in *Timed Tests: Addition and Subtraction* provide students with opportunities to practice and memorize addition and subtraction facts. Challenge students to do their personal best by timing them as they complete each test and keeping track of those times. The tests are intended to be repeated by students until they have reached the desired accuracy and speed.

Have students write a goal time on each test. When they complete a test, have them write their actual time on their test. Encourage students to retake a test to try to better their time and accuracy. Give students record sheets (pages 4 and 63) to keep track of their own scores. An answer key is provided at the back of the book for quick reference.

For extra practice, have students complete the challenge pages. The addition and subtraction challenge pages include problems with missing addends or minuends.

Reward students for earning perfect scores, meeting their personal time goals, or for mastering all the addition and subtraction facts.

Use these motivating timed tests to "recharge" skill review and give students the power to succeed!

Name _____

Addition Record Sheet

Test	Time	Score	Test	Time	Score
Warm-Up +2			Challenge +11		
Practice +2			Warm-Up +12		
Challenge +2			Practice +12		
Warm-Up +3			Challenge +12		
Practice +3			Warm-Up +13		
Challenge +3			Practice +13		
Warm-Up +4			Challenge +13		
Practice +4			Warm-Up +14		
Challenge +4			Practice +14		
Warm-Up +5			Challenge +14		
Practice +5			Warm-Up +15		
Challenge +5			Practice +15		
Warm-Up +6			Challenge +15		
Practice +6			Warm-Up +16		
Challenge +6			Practice +16		
Warm-Up +7			Challenge +16		
Practice +7			Warm-Up +17		
Challenge +7			Practice +17		
Warm-Up +8			Challenge +17		
Practice +8			Warm-Up +18		
Challenge +8			Practice +18		
Warm-Up +9			Challenge +18		
Practice +9			Warm-Up +19		
Challenge +9			Practice +19		
Warm-Up +10			Challenge +19		
Practice +10			Review +1–10		
Challenge +10			Review +11–20		
Warm-Up +11			Review +1–20A		
Practice +11			Review +1–20B		

Timed Tests: Addition and Subtraction © 2004 Creative Teaching Press

Name _____   Date _____

Goal Time _____   Actual Time _____   Score _____

 # Addition Warm-Up   +2

1 8
    + 2

2 3
    + 2

3 6
    + 2

4 4
    + 2

5 14
    + 2

6 10
    + 2

7 1
    + 2

8 2
    + 2

9 19
    + 2

10 5
    + 2

11 13
    + 2

12 9
    + 2

13 15
    + 2

14 11
    + 2

15 17
    + 2

16 12
    + 2

17 18
    + 2

18 7
    + 2

19 20
    + 2

20 16
    + 2

Name _____ Date _____

Goal Time _____ Actual Time _____ Score _____

Addition Practice   +2

1 2
   + 5

2 2
   +10

3 2
   + 8

4 2
   +19

5 2
   +15

6 2
   + 3

7 2
   + 1

8 2
   +14

9 2
   +12

10 2
   +18

11 2
   + 6

12 2
   +17

13 2
   + 9

14 2
   + 2

15 2
   +11

16 2
   +16

17 2
   + 4

18 2
   +20

19 2
   +13

20 2
   + 7

Timed Tests: Addition and Subtraction © 2004 Creative Teaching Press

Name _____ Date _____

Goal Time _____ Actual Time _____ Score _____

 # Addition Challenge +2

1 ☐
 + 2
 12

2 ☐
 + 2
 5

3 ☐
 + 2
 18

4 ☐
 + 2
 11

5 ☐
 + 2
 16

6 ☐
 + 2
 8

7 ☐
 + 2
 19

8 ☐
 + 2
 3

9 ☐
 + 2
 14

10 ☐
 + 2
 21

11 ☐
 + 2
 9

12 ☐
 + 2
 7

13 ☐
 + 2
 13

14 ☐
 + 2
 10

15 ☐
 + 2
 4

16 ☐
 + 2
 17

17 ☐
 + 2
 20

18 ☐
 + 2
 6

19 ☐
 + 2
 15

20 ☐
 + 2
 22

Name _____ Date _____

Goal Time _____ Actual Time _____ Score _____

 # Addition Warm-Up +3

1 3
    + 3

2 13
    + 3

3 6
    + 3

4 1
    + 3

5 14
    + 3

6 10
    + 3

7 7
    + 3

8 4
    + 3

9 19
    + 3

10 9
    + 3

11 5
    + 3

12 8
    + 3

13 16
    + 3

14 11
    + 3

15 17
    + 3

16 12
    + 3

17 20
    + 3

18 18
    + 3

19 2
    + 3

20 15
    + 3

Timed Tests: Addition and Subtraction © 2004 Creative Teaching Press

Goal Time _____ Actual Time _____ Score _____

Addition Practice   +3

1 $\begin{array}{r} 3 \\ +\ 9 \\ \hline \end{array}$ **2** $\begin{array}{r} 3 \\ +10 \\ \hline \end{array}$ **3** $\begin{array}{r} 3 \\ +\ 3 \\ \hline \end{array}$ **4** $\begin{array}{r} 3 \\ +19 \\ \hline \end{array}$ **5** $\begin{array}{r} 3 \\ +16 \\ \hline \end{array}$

6 $\begin{array}{r} 3 \\ +14 \\ \hline \end{array}$ **7** $\begin{array}{r} 3 \\ +15 \\ \hline \end{array}$ **8** $\begin{array}{r} 3 \\ +13 \\ \hline \end{array}$ **9** $\begin{array}{r} 3 \\ +12 \\ \hline \end{array}$ **10** $\begin{array}{r} 3 \\ +18 \\ \hline \end{array}$

11 $\begin{array}{r} 3 \\ +\ 8 \\ \hline \end{array}$ **12** $\begin{array}{r} 3 \\ +\ 5 \\ \hline \end{array}$ **13** $\begin{array}{r} 3 \\ +17 \\ \hline \end{array}$ **14** $\begin{array}{r} 3 \\ +\ 1 \\ \hline \end{array}$ **15** $\begin{array}{r} 3 \\ +\ 6 \\ \hline \end{array}$

16 $\begin{array}{r} 3 \\ +11 \\ \hline \end{array}$ **17** $\begin{array}{r} 3 \\ +\ 2 \\ \hline \end{array}$ **18** $\begin{array}{r} 3 \\ +20 \\ \hline \end{array}$ **19** $\begin{array}{r} 3 \\ +\ 7 \\ \hline \end{array}$ **20** $\begin{array}{r} 3 \\ +\ 4 \\ \hline \end{array}$

Name _____ Date _____

Goal Time _____ Actual Time _____ Score _____

Addition Challenge   +3

①
```
  ☐
+ 3
 18
```

②
```
  ☐
+ 3
 10
```

③
```
  ☐
+ 3
 22
```

④
```
  ☐
+ 3
  7
```

⑤
```
  ☐
+ 3
 14
```

⑥
```
  ☐
+ 3
 12
```

⑦
```
  ☐
+ 3
 19
```

⑧
```
  ☐
+ 3
  4
```

⑨
```
  ☐
+ 3
 16
```

⑩
```
  ☐
+ 3
 21
```

⑪
```
  ☐
+ 3
  6
```

⑫
```
  ☐
+ 3
 13
```

⑬
```
  ☐
+ 3
 23
```

⑭
```
  ☐
+ 3
  9
```

⑮
```
  ☐
+ 3
 17
```

⑯
```
  ☐
+ 3
 15
```

⑰
```
  ☐
+ 3
  8
```

⑱
```
  ☐
+ 3
 11
```

⑲
```
  ☐
+ 3
 20
```

⑳
```
  ☐
+ 3
  5
```

Timed Tests: Addition and Subtraction © 2004 Creative Teaching Press

Addition Warm-Up +4

1 15
    + 4

2   3
    + 4

3 12
    + 4

4 19
    + 4

5   6
    + 4

6   2
    + 4

7 17
    + 4

8 10
    + 4

9   8
    + 4

10 14
    + 4

11 13
    + 4

12   1
    + 4

13   5
    + 4

14 16
    + 4

15   9
    + 4

16 20
    + 4

17   7
    + 4

18 11
    + 4

19 18
    + 4

20   4
    + 4

Addition Practice   +4

1 $\begin{array}{r} 4 \\ +14 \\ \hline \end{array}$ **2** $\begin{array}{r} 4 \\ +\ 2 \\ \hline \end{array}$ **3** $\begin{array}{r} 4 \\ +19 \\ \hline \end{array}$ **4** $\begin{array}{r} 4 \\ +\ 8 \\ \hline \end{array}$ **5** $\begin{array}{r} 4 \\ +\ 9 \\ \hline \end{array}$

6 $\begin{array}{r} 4 \\ +\ 4 \\ \hline \end{array}$ **7** $\begin{array}{r} 4 \\ +18 \\ \hline \end{array}$ **8** $\begin{array}{r} 4 \\ +11 \\ \hline \end{array}$ **9** $\begin{array}{r} 4 \\ +\ 5 \\ \hline \end{array}$ **10** $\begin{array}{r} 4 \\ +15 \\ \hline \end{array}$

11 $\begin{array}{r} 4 \\ +10 \\ \hline \end{array}$ **12** $\begin{array}{r} 4 \\ +\ 6 \\ \hline \end{array}$ **13** $\begin{array}{r} 4 \\ +17 \\ \hline \end{array}$ **14** $\begin{array}{r} 4 \\ +12 \\ \hline \end{array}$ **15** $\begin{array}{r} 4 \\ +\ 1 \\ \hline \end{array}$

16 $\begin{array}{r} 4 \\ +13 \\ \hline \end{array}$ **17** $\begin{array}{r} 4 \\ +20 \\ \hline \end{array}$ **18** $\begin{array}{r} 4 \\ +16 \\ \hline \end{array}$ **19** $\begin{array}{r} 4 \\ +\ 3 \\ \hline \end{array}$ **20** $\begin{array}{r} 4 \\ +\ 7 \\ \hline \end{array}$

Timed Tests: Addition and Subtraction © 2004 Creative Teaching Press

Name _____   Date _____

Goal Time _____   Actual Time _____   Score _____

Addition Challenge   +4

1 ☐
  + 4
  ___
   7

2 ☐
  + 4
  ___
   21

3 ☐
  + 4
  ___
   15

4 ☐
  + 4
  ___
   11

5 ☐
  + 4
  ___
   19

6 ☐
  + 4
  ___
   16

7 ☐
  + 4
  ___
   13

8 ☐
  + 4
  ___
   24

9 ☐
  + 4
  ___
   9

10 ☐
  + 4
  ___
   22

11 ☐
  + 4
  ___
   10

12 ☐
  + 4
  ___
   18

13 ☐
  + 4
  ___
   5

14 ☐
  + 4
  ___
   20

15 ☐
  + 4
  ___
   14

16 ☐
  + 4
  ___
   17

17 ☐
  + 4
  ___
   8

18 ☐
  + 4
  ___
   23

19 ☐
  + 4
  ___
   12

20 ☐
  + 4
  ___
   6

Addition Warm-Up   +5

1 1
    + 5

2 15
    + 5

3 11
    + 5

4 7
    + 5

5 18
    + 5

6 17
    + 5

7 3
    + 5

8 19
    + 5

9 13
    + 5

10 8
    + 5

11 14
    + 5

12 9
    + 5

13 5
    + 5

14 16
    + 5

15 2
    + 5

16 4
    + 5

17 20
    + 5

18 12
    + 5

19 6
    + 5

20 10
    + 5

Timed Tests: Addition and Subtraction © 2004 Creative Teaching Press

Name _____ Date _____

Goal Time _____ Actual Time _____ Score _____

Addition Practice   +5

1 $\begin{array}{r} 5 \\ +14 \\ \hline \end{array}$ **2** $\begin{array}{r} 5 \\ +\ 2 \\ \hline \end{array}$ **3** $\begin{array}{r} 5 \\ +20 \\ \hline \end{array}$ **4** $\begin{array}{r} 5 \\ +\ 8 \\ \hline \end{array}$ **5** $\begin{array}{r} 5 \\ +16 \\ \hline \end{array}$

6 $\begin{array}{r} 5 \\ +18 \\ \hline \end{array}$ **7** $\begin{array}{r} 5 \\ +\ 4 \\ \hline \end{array}$ **8** $\begin{array}{r} 5 \\ +13 \\ \hline \end{array}$ **9** $\begin{array}{r} 5 \\ +\ 6 \\ \hline \end{array}$ **10** $\begin{array}{r} 5 \\ +10 \\ \hline \end{array}$

11 $\begin{array}{r} 5 \\ +11 \\ \hline \end{array}$ **12** $\begin{array}{r} 5 \\ +\ 7 \\ \hline \end{array}$ **13** $\begin{array}{r} 5 \\ +15 \\ \hline \end{array}$ **14** $\begin{array}{r} 5 \\ +\ 1 \\ \hline \end{array}$ **15** $\begin{array}{r} 5 \\ +12 \\ \hline \end{array}$

16 $\begin{array}{r} 5 \\ +19 \\ \hline \end{array}$ **17** $\begin{array}{r} 5 \\ +\ 3 \\ \hline \end{array}$ **18** $\begin{array}{r} 5 \\ +\ 9 \\ \hline \end{array}$ **19** $\begin{array}{r} 5 \\ +17 \\ \hline \end{array}$ **20** $\begin{array}{r} 5 \\ +\ 5 \\ \hline \end{array}$

Name _____   Date _____

Goal Time _____   Actual Time _____   Score _____

Addition Challenge   +5

1 ☐
$\begin{array}{r} +\ 5 \\ \hline 10 \end{array}$

2 ☐
$\begin{array}{r} +\ 5 \\ \hline 21 \end{array}$

3 ☐
$\begin{array}{r} +\ 5 \\ \hline 8 \end{array}$

4 ☐
$\begin{array}{r} +\ 5 \\ \hline 24 \end{array}$

5 ☐
$\begin{array}{r} +\ 5 \\ \hline 15 \end{array}$

6 ☐
$\begin{array}{r} +\ 5 \\ \hline 18 \end{array}$

7 ☐
$\begin{array}{r} +\ 5 \\ \hline 23 \end{array}$

8 ☐
$\begin{array}{r} +\ 5 \\ \hline 12 \end{array}$

9 ☐
$\begin{array}{r} +\ 5 \\ \hline 25 \end{array}$

10 ☐
$\begin{array}{r} +\ 5 \\ \hline 6 \end{array}$

11 ☐
$\begin{array}{r} +\ 5 \\ \hline 16 \end{array}$

12 ☐
$\begin{array}{r} +\ 5 \\ \hline 7 \end{array}$

13 ☐
$\begin{array}{r} +\ 5 \\ \hline 19 \end{array}$

14 ☐
$\begin{array}{r} +\ 5 \\ \hline 22 \end{array}$

15 ☐
$\begin{array}{r} +\ 5 \\ \hline 13 \end{array}$

16 ☐
$\begin{array}{r} +\ 5 \\ \hline 9 \end{array}$

17 ☐
$\begin{array}{r} +\ 5 \\ \hline 14 \end{array}$

18 ☐
$\begin{array}{r} +\ 5 \\ \hline 20 \end{array}$

19 ☐
$\begin{array}{r} +\ 5 \\ \hline 11 \end{array}$

20 ☐
$\begin{array}{r} +\ 5 \\ \hline 17 \end{array}$

Timed Tests: Addition and Subtraction © 2004 Creative Teaching Press

Goal Time _____   Actual Time _____   Score _____

 Addition Warm-Up   +6

1 9
   + 6

2 18
   + 6

3 2
   + 6

4 14
   + 6

5 8
   + 6

6 20
   + 6

7 1
   + 6

8 16
   + 6

9 5
   + 6

10 12
    + 6

11 4
    + 6

12 13
    + 6

13 7
    + 6

14 19
    + 6

15 10
    + 6

16 17
    + 6

17 6
    + 6

18 11
    + 6

19 15
    + 6

20 3
    + 6

Name _____ Date _____

Goal Time _____ Actual Time _____ Score _____

 # Addition Practice   +6

① 6
+14

② 6
+18

③ 6
+ 3

④ 6
+10

⑤ 6
+ 6

⑥ 6
+ 1

⑦ 6
+12

⑧ 6
+ 5

⑨ 6
+ 8

⑩ 6
+17

⑪ 6
+16

⑫ 6
+19

⑬ 6
+ 2

⑭ 6
+ 7

⑮ 6
+11

⑯ 6
+ 4

⑰ 6
+15

⑱ 6
+20

⑲ 6
+ 9

⑳ 6
+13

Timed Tests: Addition and Subtraction © 2004 Creative Teaching Press

Goal Time _____   Actual Time _____   Score _____

Addition Challenge   +6

1 ☐
　+ 6
　 7

2 ☐
　+ 6
　 18

3 ☐
　+ 6
　 12

4 ☐
　+ 6
　 26

5 ☐
　+ 6
　 15

6 ☐
　+ 6
　 13

7 ☐
　+ 6
　 21

8 ☐
　+ 6
　 17

9 ☐
　+ 6
　 10

10 ☐
　+ 6
　 23

11 ☐
　+ 6
　 24

12 ☐
　+ 6
　 9

13 ☐
　+ 6
　 22

14 ☐
　+ 6
　 16

15 ☐
　+ 6
　 8

16 ☐
　+ 6
　 11

17 ☐
　+ 6
　 19

18 ☐
　+ 6
　 25

19 ☐
　+ 6
　 14

20 ☐
　+ 6
　 20

Name _____ Date _____

Goal Time _____ Actual Time _____ Score _____

 # Addition Warm-Up   +7

1 2
   + 7

2 10
   + 7

3 19
   + 7

4 5
   + 7

5 13
   + 7

6 14
   + 7

7 6
   + 7

8 1
   + 7

9 17
   + 7

10 8
   + 7

11 7
   + 7

12 18
   + 7

13 9
   + 7

14 11
   + 7

15 3
   + 7

16 16
   + 7

17 4
   + 7

18 12
   + 7

19 20
   + 7

20 15
   + 7

Timed Tests: Addition and Subtraction © 2004 Creative Teaching Press

Name _____ Date _____

Goal Time _____ Actual Time _____ Score _____

Addition Practice   +7

1   7
    + 9

2   7
    +13

3   7
    + 5

4   7
    +16

5   7
    +11

6   7
    + 6

7   7
    +17

8   7
    + 8

9   7
    + 2

10   7
    +20

11   7
    +18

12   7
    + 3

13   7
    +10

14   7
    +15

15   7
    + 4

16   7
    + 1

17   7
    +14

18   7
    + 7

19   7
    +19

20   7
    +12

Goal Time _____ Actual Time _____ Score _____

Addition Challenge   +7

1 ☐
+ 7
22

2 ☐
+ 7
13

3 ☐
+ 7
25

4 ☐
+ 7
18

5 ☐
+ 7
10

6 ☐
+ 7
16

7 ☐
+ 7
8

8 ☐
+ 7
20

9 ☐
+ 7
12

10 ☐
+ 7
26

11 ☐
+ 7
24

12 ☐
+ 7
17

13 ☐
+ 7
14

14 ☐
+ 7
9

15 ☐
+ 7
21

16 ☐
+ 7
11

17 ☐
+ 7
27

18 ☐
+ 7
19

19 ☐
+ 7
23

20 ☐
+ 7
15

Timed Tests: Addition and Subtraction © 2004 Creative Teaching Press

Name _____ Date _____

Goal Time _____ Actual Time _____ Score _____

Addition Warm-Up +8

① 7
+ 8

② 12
+ 8

③ 9
+ 8

④ 16
+ 8

⑤ 1
+ 8

⑥ 19
+ 8

⑦ 2
+ 8

⑧ 14
+ 8

⑨ 6
+ 8

⑩ 17
+ 8

⑪ 4
+ 8

⑫ 10
+ 8

⑬ 18
+ 8

⑭ 3
+ 8

⑮ 15
+ 8

⑯ 13
+ 8

⑰ 5
+ 8

⑱ 20
+ 8

⑲ 8
+ 8

⑳ 11
+ 8

Name _____   Date _____

Goal Time _____   Actual Time _____   Score _____

 # Addition Practice   +8

① 8
   + 2

② 8
   +16

③ 8
   + 8

④ 8
   + 4

⑤ 8
   +19

⑥ 8
   + 7

⑦ 8
   +11

⑧ 8
   + 5

⑨ 8
   +14

⑩ 8
   +18

⑪ 8
   + 1

⑫ 8
   + 9

⑬ 8
   +20

⑭ 8
   +12

⑮ 8
   + 6

⑯ 8
   +13

⑰ 8
   +17

⑱ 8
   + 3

⑲ 8
   +15

⑳ 8
   +10

Timed Tests: Addition and Subtraction © 2004 Creative Teaching Press

Name _____ Date _____

Goal Time _____ Actual Time _____ Score _____

 # Addition Challenge   +8

1 ☐
+ 8
16

2 ☐
+ 8
23

3 ☐
+ 8
10

4 ☐
+ 8
26

5 ☐
+ 8
15

6 ☐
+ 8
18

7 ☐
+ 8
14

8 ☐
+ 8
27

9 ☐
+ 8
12

10 ☐
+ 8
19

11 ☐
+ 8
21

12 ☐
+ 8
9

13 ☐
+ 8
24

14 ☐
+ 8
17

15 ☐
+ 8
28

16 ☐
+ 8
25

17 ☐
+ 8
13

18 ☐
+ 8
20

19 ☐
+ 8
11

20 ☐
+ 8
22

Name _____ Date _____

Goal Time _____ Actual Time _____ Score _____

 # Addition Warm-Up +9

1 10 +9

2 13 +9

3 4 +9

4 18 +9

5 6 +9

6 7 +9

7 16 +9

8 1 +9

9 9 +9

10 14 +9

11 2 +9

12 19 +9

13 11 +9

14 5 +9

15 17 +9

16 8 +9

17 15 +9

18 3 +9

19 20 +9

20 12 +9

26

Timed Tests: Addition and Subtraction © 2004 Creative Teaching Press

Goal Time _____ Actual Time _____ Score _____

Addition Practice   +9

1 $\begin{array}{r} 9 \\ +20 \\ \hline \end{array}$ **2** $\begin{array}{r} 9 \\ +\ 8 \\ \hline \end{array}$ **3** $\begin{array}{r} 9 \\ +16 \\ \hline \end{array}$ **4** $\begin{array}{r} 9 \\ +\ 3 \\ \hline \end{array}$ **5** $\begin{array}{r} 9 \\ +19 \\ \hline \end{array}$

6 $\begin{array}{r} 9 \\ +12 \\ \hline \end{array}$ **7** $\begin{array}{r} 9 \\ +14 \\ \hline \end{array}$ **8** $\begin{array}{r} 9 \\ +\ 5 \\ \hline \end{array}$ **9** $\begin{array}{r} 9 \\ +10 \\ \hline \end{array}$ **10** $\begin{array}{r} 9 \\ +17 \\ \hline \end{array}$

11 $\begin{array}{r} 9 \\ +18 \\ \hline \end{array}$ **12** $\begin{array}{r} 9 \\ +\ 2 \\ \hline \end{array}$ **13** $\begin{array}{r} 9 \\ +\ 9 \\ \hline \end{array}$ **14** $\begin{array}{r} 9 \\ +\ 6 \\ \hline \end{array}$ **15** $\begin{array}{r} 9 \\ +13 \\ \hline \end{array}$

16 $\begin{array}{r} 9 \\ +\ 4 \\ \hline \end{array}$ **17** $\begin{array}{r} 9 \\ +11 \\ \hline \end{array}$ **18** $\begin{array}{r} 9 \\ +\ 7 \\ \hline \end{array}$ **19** $\begin{array}{r} 9 \\ +15 \\ \hline \end{array}$ **20** $\begin{array}{r} 9 \\ +\ 1 \\ \hline \end{array}$

Addition Challenge +9

① $\boxed{}$
$$\begin{array}{r} +\ 9 \\ \hline 14 \end{array}$$

② $\boxed{}$
$$\begin{array}{r} +\ 9 \\ \hline 22 \end{array}$$

③ $\boxed{}$
$$\begin{array}{r} +\ 9 \\ \hline 20 \end{array}$$

④ $\boxed{}$
$$\begin{array}{r} +\ 9 \\ \hline 11 \end{array}$$

⑤ $\boxed{}$
$$\begin{array}{r} +\ 9 \\ \hline 27 \end{array}$$

⑥ $\boxed{}$
$$\begin{array}{r} +\ 9 \\ \hline 16 \end{array}$$

⑦ $\boxed{}$
$$\begin{array}{r} +\ 9 \\ \hline 24 \end{array}$$

⑧ $\boxed{}$
$$\begin{array}{r} +\ 9 \\ \hline 12 \end{array}$$

⑨ $\boxed{}$
$$\begin{array}{r} +\ 9 \\ \hline 29 \end{array}$$

⑩ $\boxed{}$
$$\begin{array}{r} +\ 9 \\ \hline 18 \end{array}$$

⑪ $\boxed{}$
$$\begin{array}{r} +\ 9 \\ \hline 25 \end{array}$$

⑫ $\boxed{}$
$$\begin{array}{r} +\ 9 \\ \hline 10 \end{array}$$

⑬ $\boxed{}$
$$\begin{array}{r} +\ 9 \\ \hline 21 \end{array}$$

⑭ $\boxed{}$
$$\begin{array}{r} +\ 9 \\ \hline 15 \end{array}$$

⑮ $\boxed{}$
$$\begin{array}{r} +\ 9 \\ \hline 23 \end{array}$$

⑯ $\boxed{}$
$$\begin{array}{r} +\ 9 \\ \hline 19 \end{array}$$

⑰ $\boxed{}$
$$\begin{array}{r} +\ 9 \\ \hline 17 \end{array}$$

⑱ $\boxed{}$
$$\begin{array}{r} +\ 9 \\ \hline 26 \end{array}$$

⑲ $\boxed{}$
$$\begin{array}{r} +\ 9 \\ \hline 13 \end{array}$$

⑳ $\boxed{}$
$$\begin{array}{r} +\ 9 \\ \hline 28 \end{array}$$

Name _____ Date _____

Goal Time _____ Actual Time _____ Score _____

Addition Warm-Up +10

1 3
   +10

2 17
   +10

3 8
   +10

4 20
   +10

5 13
   +10

6 6
   +10

7 4
   +10

8 18
   +10

9 12
   +10

10 7
   +10

11 14
   +10

12 10
   +10

13 2
   +10

14 5
   +10

15 19
   +10

16 1
   +10

17 11
   +10

18 9
   +10

19 16
   +10

20 15
   +10

Name _____ Date _____

Goal Time _____ Actual Time _____ Score _____

 # Addition Practice   +10

1 10
  + 9

2 10
  + 7

3 10
  +12

4 10
  +18

5 10
  + 3

6 10
  + 4

7 10
  +19

8 10
  + 1

9 10
  +15

10 10
  +13

11 10
  +14

12 10
  + 2

13 10
  +11

14 10
  + 8

15 10
  +16

16 10
  + 5

17 10
  +10

18 10
  + 6

19 10
  +17

20 10
  +20

Timed Tests: Addition and Subtraction © 2004 Creative Teaching Press

Addition Challenge +10

1 ☐
+10
21

2 ☐
+10
17

3 ☐
+10
28

4 ☐
+10
15

5 ☐
+10
23

6 ☐
+10
14

7 ☐
+10
29

8 ☐
+10
12

9 ☐
+10
25

10 ☐
+10
11

11 ☐
+10
19

12 ☐
+10
24

13 ☐
+10
18

14 ☐
+10
30

15 ☐
+10
26

16 ☐
+10
22

17 ☐
+10
16

18 ☐
+10
27

19 ☐
+10
13

20 ☐
+10
20

Name _____ Date _____

Goal Time _____ Actual Time _____ Score _____

 # Addition Warm-Up   +11

① 6
+11

② 20
+11

③ 4
+11

④ 17
+11

⑤ 14
+11

⑥ 19
+11

⑦ 8
+11

⑧ 1
+11

⑨ 12
+11

⑩ 10
+11

⑪ 2
+11

⑫ 15
+11

⑬ 9
+11

⑭ 5
+11

⑮ 13
+11

⑯ 11
+11

⑰ 3
+11

⑱ 18
+11

⑲ 7
+11

⑳ 16
+11

Goal Time _____ Actual Time _____ Score _____

Addition Practice +11

1 11
  + 5

2 11
  +10

3 11
  + 8

4 11
  +19

5 11
  +13

6 11
  +16

7 11
  + 3

8 11
  +14

9 11
  + 6

10 11
  +11

11 11
  + 4

12 11
  + 9

13 11
  +12

14 11
  + 1

15 11
  +20

16 11
  +18

17 11
  + 2

18 11
  +15

19 11
  + 7

20 11
  +17

Goal Time _____ Actual Time _____ Score _____

Addition Challenge   +11

1 ▢
$\begin{array}{r} +11 \\ \hline 22 \end{array}$

2 ▢
$\begin{array}{r} +11 \\ \hline 30 \end{array}$

3 ▢
$\begin{array}{r} +11 \\ \hline 12 \end{array}$

4 ▢
$\begin{array}{r} +11 \\ \hline 25 \end{array}$

5 ▢
$\begin{array}{r} +11 \\ \hline 19 \end{array}$

6 ▢
$\begin{array}{r} +11 \\ \hline 16 \end{array}$

7 ▢
$\begin{array}{r} +11 \\ \hline 26 \end{array}$

8 ▢
$\begin{array}{r} +11 \\ \hline 18 \end{array}$

9 ▢
$\begin{array}{r} +11 \\ \hline 31 \end{array}$

10 ▢
$\begin{array}{r} +11 \\ \hline 15 \end{array}$

11 ▢
$\begin{array}{r} +11 \\ \hline 29 \end{array}$

12 ▢
$\begin{array}{r} +11 \\ \hline 14 \end{array}$

13 ▢
$\begin{array}{r} +11 \\ \hline 21 \end{array}$

14 ▢
$\begin{array}{r} +11 \\ \hline 27 \end{array}$

15 ▢
$\begin{array}{r} +11 \\ \hline 23 \end{array}$

16 ▢
$\begin{array}{r} +11 \\ \hline 20 \end{array}$

17 ▢
$\begin{array}{r} +11 \\ \hline 17 \end{array}$

18 ▢
$\begin{array}{r} +11 \\ \hline 24 \end{array}$

19 ▢
$\begin{array}{r} +11 \\ \hline 13 \end{array}$

20 ▢
$\begin{array}{r} +11 \\ \hline 28 \end{array}$

Timed Tests: Addition and Subtraction © 2004 Creative Teaching Press

Goal Time _____ Actual Time _____ Score _____

 # Addition Warm-Up +12

1 3
  +12

2 11
  +12

3 17
  +12

4 6
  +12

5 14
  +12

6 19
  +12

7 7
  +12

8 2
  +12

9 15
  +12

10 20
  +12

11 5
  +12

12 12
  +12

13 9
  +12

14 18
  +12

15 10
  +12

16 1
  +12

17 8
  +12

18 16
  +12

19 13
  +12

20 4
  +12

Goal Time _____   Actual Time _____   Score _____

 Addition Practice   +12

1 12
  +18

2 12
  + 8

3 12
  +13

4 12
  + 4

5 12
  +20

6 12
  + 5

7 12
  +12

8 12
  +10

9 12
  +16

10 12
  + 1

11 12
  +14

12 12
  + 2

13 12
  +17

14 12
  + 6

15 12
  +19

16 12
  + 7

17 12
  +11

18 12
  + 3

19 12
  +15

20 12
  + 9

Name _____ Date _____

Goal Time _____ Actual Time _____ Score _____

Addition Challenge +12

1 ☐
+12
24

2 ☐
+12
31

3 ☐
+12
13

4 ☐
+12
27

5 ☐
+12
17

6 ☐
+12
14

7 ☐
+12
23

8 ☐
+12
18

9 ☐
+12
29

10 ☐
+12
20

11 ☐
+12
22

12 ☐
+12
28

13 ☐
+12
32

14 ☐
+12
25

15 ☐
+12
15

16 ☐
+12
26

17 ☐
+12
21

18 ☐
+12
16

19 ☐
+12
30

20 ☐
+12
19

Timed Tests: Addition and Subtraction © 2004 Creative Teaching Press

Name _____ Date _____

Goal Time _____ Actual Time _____ Score _____

 # Addition Warm-Up +13

1   6
    +13

2   19
    +13

3   5
    +13

4   11
    +13

5   16
    +13

6   13
    +13

7   4
    +13

8   20
    +13

9   9
    +13

10   2
    +13

11   8
    +13

12   15
    +13

13   1
    +13

14   14
    +13

15   7
    +13

16   3
    +13

17   10
    +13

18   17
    +13

19   12
    +13

20   18
    +13

Timed Tests: Addition and Subtraction © 2004 Creative Teaching Press

Name _____ Date _____

Goal Time _____ Actual Time _____ Score _____

 Addition Practice   +13

1 13
    +11

2 13
    + 3

3 13
    +16

4 13
    +20

5 13
    + 6

6 13
    + 5

7 13
    + 8

8 13
    +17

9 13
    + 1

10 13
     +12

11 13
     +10

12 13
     + 4

13 13
     +19

14 13
     + 7

15 13
     + 9

16 13
     +14

17 13
     + 2

18 13
     +18

19 13
     +15

20 13
     +13

Name _____ Date _____

Goal Time _____ Actual Time _____ Score _____

 # Addition Challenge   +13

① ☐
   +13
   14

② ☐
   +13
   22

③ ☐
   +13
   30

④ ☐
   +13
   18

⑤ ☐
   +13
   26

⑥ ☐
   +13
   28

⑦ ☐
   +13
   17

⑧ ☐
   +13
   32

⑨ ☐
   +13
   21

⑩ ☐
   +13
   24

⑪ ☐
   +13
   25

⑫ ☐
   +13
   20

⑬ ☐
   +13
   15

⑭ ☐
   +13
   27

⑮ ☐
   +13
   31

⑯ ☐
   +13
   16

⑰ ☐
   +13
   29

⑱ ☐
   +13
   23

⑲ ☐
   +13
   33

⑳ ☐
   +13
   19

Timed Tests: Addition and Subtraction © 2004 Creative Teaching Press

Name _____ Date _____

Goal Time _____ Actual Time _____ Score _____

 # Addition Warm-Up +14

1 12
 +14

2   6
 +14

3 19
 +14

4   2
 +14

5 15
 +14

6   4
 +14

7 20
 +14

8   7
 +14

9 18
 +14

10 10
 +14

11 17
 +14

12   8
 +14

13 11
 +14

14   1
 +14

15 13
 +14

16   9
 +14

17   3
 +14

18 14
 +14

19   5
 +14

20 16
 +14

Timed Tests: Addition and Subtraction © 2004 Creative Teaching Press

Name _____ Date _____

Goal Time _____ Actual Time _____ Score _____

Addition Practice   +14

1 14
     + 9

2 14
     +11

3 14
     + 8

4 14
     + 3

5 14
     +18

6 14
     + 1

7 14
     +16

8 14
     + 5

9 14
     +12

10 14
      +19

11 14
      + 7

12 14
      +13

13 14
      + 2

14 14
      + 6

15 14
      +10

16 14
      + 4

17 14
      +14

18 14
      +15

19 14
      +20

20 14
      +17

42

Timed Tests: Addition and Subtraction © 2004 Creative Teaching Press

Name _____ Date _____

Goal Time _____ Actual Time _____ Score _____

 # Addition Challenge +14

1 ☐
+14
30

2 ☐
+14
17

3 ☐
+14
28

4 ☐
+14
20

5 ☐
+14
25

6 ☐
+14
19

7 ☐
+14
23

8 ☐
+14
31

9 ☐
+14
15

10 ☐
+14
33

11 ☐
+14
27

12 ☐
+14
18

13 ☐
+14
24

14 ☐
+14
32

15 ☐
+14
21

16 ☐
+14
26

17 ☐
+14
34

18 ☐
+14
16

19 ☐
+14
22

20 ☐
+14
29

Name _____ Date _____

Goal Time _____ Actual Time _____ Score _____

 # Addition Warm-Up   +15

1 5
      +15

2 16
      +15

3 3
      +15

4 8
      +15

5 10
      +15

6 13
      +15

7 6
      +15

8 18
      +15

9 20
      +15

10 14
       +15

11 11
       +15

12 2
       +15

13 15
       +15

14 9
       +15

15 17
       +15

16 1
       +15

17 19
       +15

18 7
       +15

19 4
       +15

20 12
       +15

44

Timed Tests: Addition and Subtraction © 2004 Creative Teaching Press

Addition Practice   +15

1   15
     + 9

2   15
     + 3

3   15
     +14

4   15
     + 1

5   15
     + 8

6   15
     +12

7   15
     + 7

8   15
     +16

9   15
     + 5

10   15
      +19

11   15
      + 6

12   15
      +10

13   15
      + 2

14   15
      +13

15   15
      +17

16   15
      +15

17   15
      + 4

18   15
      +18

19   15
      +11

20   15
      +20

Name _____ Date _____

Goal Time _____ Actual Time _____ Score _____

 # Addition Challenge +15

❶ ☐
+15
26

❷ ☐
+15
22

❸ ☐
+15
31

❹ ☐
+15
18

❺ ☐
+15
29

❻ ☐
+15
17

❼ ☐
+15
32

❽ ☐
+15
20

❾ ☐
+15
35

❿ ☐
+15
24

⓫ ☐
+15
21

⓬ ☐
+15
30

⓭ ☐
+15
25

⓮ ☐
+15
16

⓯ ☐
+15
27

⓰ ☐
+15
28

⓱ ☐
+15
19

⓲ ☐
+15
34

⓳ ☐
+15
23

⓴ ☐
+15
33

Timed Tests: Addition and Subtraction © 2004 Creative Teaching Press

Name _____ Date _____

Goal Time _____ Actual Time _____ Score _____

 # Addition Warm-Up +16

1   1
    +16

2   14
    +16

3   7
    +16

4   18
    +16

5   5
    +16

6   12
    +16

7   9
    +16

8   4
    +16

9   15
    +16

10   20
    +16

11   6
    +16

12   17
    +16

13   8
    +16

14   10
    +16

15   2
    +16

16   3
    +16

17   16
    +16

18   11
    +16

19   13
    +16

20   19
    +16

Name _____   Date _____

Goal Time _____   Actual Time _____   Score _____

 Addition Practice   +16

1 16
     + 9

2 16
     + 3

3 16
     +16

4 16
     +18

5 16
     + 5

6 16
     + 8

7 16
     + 6

8 16
     +12

9 16
     + 4

10 16
      +10

11 16
      +14

12 16
      + 1

13 16
      +17

14 16
      + 7

15 16
      +13

16 16
      +11

17 16
      +20

18 16
      +15

19 16
      + 2

20 16
      +19

48

Timed Tests: Addition and Subtraction © 2004 Creative Teaching Press

Addition Challenge +16

1 ☐
+16
30

2 ☐
+16
32

3 ☐
+16
19

4 ☐
+16
35

5 ☐
+16
26

6 ☐
+16
20

7 ☐
+16
27

8 ☐
+16
36

9 ☐
+16
23

10 ☐
+16
31

11 ☐
+16
24

12 ☐
+16
18

13 ☐
+16
34

14 ☐
+16
21

15 ☐
+16
28

16 ☐
+16
29

17 ☐
+16
22

18 ☐
+16
33

19 ☐
+16
25

20 ☐
+16
17

Name _____ Date _____

Goal Time _____ Actual Time _____ Score _____

 # Addition Warm-Up   +17

1 5
  +17

2 15
  +17

3 10
  +17

4 8
  +17

5 13
  +17

6 16
  +17

7 4
  +17

8 19
  +17

9 2
  +17

10 20
  +17

11 9
  +17

12 1
  +17

13 18
  +17

14 6
  +17

15 11
  +17

16 3
  +17

17 12
  +17

18 7
  +17

19 14
  +17

20 17
  +17

Name _____ Date _____

Goal Time _____ Actual Time _____ Score _____

 # Addition Practice   +17

1 17
+ 2

2 17
+13

3 17
+20

4 17
+ 4

5 17
+17

6 17
+ 6

7 17
+ 9

8 17
+ 1

9 17
+19

10 17
+14

11 17
+12

12 17
+ 7

13 17
+ 3

14 17
+15

15 17
+10

16 17
+16

17 17
+ 5

18 17
+11

19 17
+ 8

20 17
+18

Timed Tests: Addition and Subtraction © 2004 Creative Teaching Press

Name _____ Date _____

Goal Time _____ Actual Time _____ Score _____

 Addition Challenge   +17

❶ ☐
+17
20

❷ ☐
+17
26

❸ ☐
+17
34

❹ ☐
+17
23

❺ ☐
+17
31

❻ ☐
+17
28

❼ ☐
+17
22

❽ ☐
+17
35

❾ ☐
+17
30

❿ ☐
+17
18

⓫ ☐
+17
32

⓬ ☐
+17
19

⓭ ☐
+17
37

⓮ ☐
+17
25

⓯ ☐
+17
36

⓰ ☐
+17
24

⓱ ☐
+17
27

⓲ ☐
+17
33

⓳ ☐
+17
21

⓴ ☐
+17
29

Timed Tests: Addition and Subtraction © 2004 Creative Teaching Press

Addition Warm-Up +18

1 9
    +18

2 5
    +18

3 12
    +18

4 1
    +18

5 16
    +18

6 14
    +18

7 2
    +18

8 20
    +18

9 7
    +18

10 18
     +18

11 4
     +18

12 17
     +18

13 8
     +18

14 13
     +18

15 10
     +18

16 19
     +18

17 6
     +18

18 11
     +18

19 3
     +18

20 15
     +18

Name _____ Date _____

Goal Time _____ Actual Time _____ Score _____

Addition Practice   +18

1 18
+ 6

2 18
+15

3 18
+ 2

4 18
+13

5 18
+19

6 18
+ 7

7 18
+14

8 18
+17

9 18
+ 3

10 18
+12

11 18
+20

12 18
+11

13 18
+16

14 18
+ 5

15 18
+18

16 18
+ 9

17 18
+ 1

18 18
+10

19 18
+ 4

20 18
+ 8

Addition Challenge   +18

1 ☐
+18
32

2 ☐
+18
21

3 ☐
+18
36

4 ☐
+18
27

5 ☐
+18
19

6 ☐
+18
23

7 ☐
+18
30

8 ☐
+18
25

9 ☐
+18
38

10 ☐
+18
33

11 ☐
+18
35

12 ☐
+18
20

13 ☐
+18
37

14 ☐
+18
24

15 ☐
+18
29

16 ☐
+18
28

17 ☐
+18
34

18 ☐
+18
22

19 ☐
+18
31

20 ☐
+18
26

Name _____ Date _____

Goal Time _____ Actual Time _____ Score _____

 # Addition Warm-Up +19

1　　5
　　+19

2　　14
　　+19

3　　7
　　+19

4　　10
　　+19

5　　2
　　+19

6　　1
　　+19

7　　20
　　+19

8　　13
　　+19

9　　9
　　+19

10　　16
　　+19

11　　12
　　+19

12　　3
　　+19

13　　18
　　+19

14　　6
　　+19

15　　11
　　+19

16　　8
　　+19

17　　15
　　+19

18　　19
　　+19

19　　4
　　+19

20　　17
　　+19

Name _____ Date _____

Goal Time _____ Actual Time _____ Score _____

Addition Practice +19

1 19
+ 3

2 19
+11

3 19
+ 8

4 19
+14

5 19
+18

6 19
+ 6

7 19
+15

8 19
+ 4

9 19
+19

10 19
+10

11 19
+ 1

12 19
+12

13 19
+20

14 19
+ 5

15 19
+16

16 19
+13

17 19
+ 9

18 19
+17

19 19
+ 2

20 19
+ 7

Goal Time _____ Actual Time _____ Score _____

Addition Challenge +19

1 ☐
+19
31

2 ☐
+19
20

3 ☐
+19
35

4 ☐
+19
24

5 ☐
+19
38

6 ☐
+19
22

7 ☐
+19
30

8 ☐
+19
28

9 ☐
+19
33

10 ☐
+19
26

11 ☐
+19
27

12 ☐
+19
34

13 ☐
+19
21

14 ☐
+19
36

15 ☐
+19
32

16 ☐
+19
37

17 ☐
+19
25

18 ☐
+19
39

19 ☐
+19
29

20 ☐
+19
23

Timed Tests: Addition and Subtraction © 2004 Creative Teaching Press

Goal Time _____ Actual Time _____ Score _____

 # Addition Review   +1–10

1 $\begin{array}{r} 2 \\ + 8 \\ \hline \end{array}$
2 $\begin{array}{r} 4 \\ + 7 \\ \hline \end{array}$
3 $\begin{array}{r} 3 \\ + 3 \\ \hline \end{array}$
4 $\begin{array}{r} 10 \\ + 4 \\ \hline \end{array}$
5 $\begin{array}{r} 6 \\ + 6 \\ \hline \end{array}$

6 $\begin{array}{r} 9 \\ + 3 \\ \hline \end{array}$
7 $\begin{array}{r} 10 \\ +10 \\ \hline \end{array}$
8 $\begin{array}{r} 3 \\ + 4 \\ \hline \end{array}$
9 $\begin{array}{r} 2 \\ + 6 \\ \hline \end{array}$
10 $\begin{array}{r} 3 \\ + 5 \\ \hline \end{array}$

11 $\begin{array}{r} 5 \\ + 4 \\ \hline \end{array}$
12 $\begin{array}{r} 4 \\ + 8 \\ \hline \end{array}$
13 $\begin{array}{r} 7 \\ + 7 \\ \hline \end{array}$
14 $\begin{array}{r} 5 \\ + 1 \\ \hline \end{array}$
15 $\begin{array}{r} 10 \\ + 8 \\ \hline \end{array}$

16 $\begin{array}{r} 2 \\ + 9 \\ \hline \end{array}$
17 $\begin{array}{r} 6 \\ + 7 \\ \hline \end{array}$
18 $\begin{array}{r} 8 \\ + 9 \\ \hline \end{array}$
19 $\begin{array}{r} 9 \\ + 6 \\ \hline \end{array}$
20 $\begin{array}{r} 4 \\ + 6 \\ \hline \end{array}$

Name _____ Date _____

Goal Time _____ Actual Time _____ Score _____

 # Addition Review   +11–20

1 19 +12   **2** 14 +16   **3** 11 +14   **4** 17 +16   **5** 15 +13

6 13 +18   **7** 16 +16   **8** 19 +17   **9** 15 +16   **10** 20 +14

11 11 +13   **12** 17 +14   **13** 12 +16   **14** 20 +15   **15** 12 +19

16 19 +18   **17** 14 +15   **18** 11 +18   **19** 13 +16   **20** 19 +16

Timed Tests: Addition and Subtraction © 2004 Creative Teaching Press

Name _____   Date _____

Addition Review   +1–20A

1 $\begin{array}{r} 6 \\ + 9 \\ \hline \end{array}$ **2** $\begin{array}{r} 10 \\ + 3 \\ \hline \end{array}$ **3** $\begin{array}{r} 20 \\ + 6 \\ \hline \end{array}$ **4** $\begin{array}{r} 8 \\ + 2 \\ \hline \end{array}$ **5** $\begin{array}{r} 13 \\ + 5 \\ \hline \end{array}$

6 $\begin{array}{r} 16 \\ + 7 \\ \hline \end{array}$ **7** $\begin{array}{r} 15 \\ + 4 \\ \hline \end{array}$ **8** $\begin{array}{r} 19 \\ +11 \\ \hline \end{array}$ **9** $\begin{array}{r} 16 \\ + 4 \\ \hline \end{array}$ **10** $\begin{array}{r} 9 \\ + 5 \\ \hline \end{array}$

11 $\begin{array}{r} 11 \\ + 5 \\ \hline \end{array}$ **12** $\begin{array}{r} 7 \\ + 7 \\ \hline \end{array}$ **13** $\begin{array}{r} 4 \\ +12 \\ \hline \end{array}$ **14** $\begin{array}{r} 17 \\ + 1 \\ \hline \end{array}$ **15** $\begin{array}{r} 3 \\ +13 \\ \hline \end{array}$

16 $\begin{array}{r} 8 \\ + 9 \\ \hline \end{array}$ **17** $\begin{array}{r} 2 \\ +20 \\ \hline \end{array}$ **18** $\begin{array}{r} 17 \\ + 8 \\ \hline \end{array}$ **19** $\begin{array}{r} 16 \\ + 1 \\ \hline \end{array}$ **20** $\begin{array}{r} 6 \\ +19 \\ \hline \end{array}$

Goal Time _____   Actual Time _____   Score _____

 Addition Review **+1–20B**

1 2
   + 9

2 8
   + 7

3 18
   + 4

4 6
   +11

5 3
   + 5

6 15
   + 8

7 12
   +10

8 13
   + 1

9 7
   +20

10 17
   + 6

11 14
   + 7

12 16
   + 2

13 19
   +19

14 4
   + 9

15 15
   + 3

16 20
   +19

17 10
   +10

18 14
   +14

19 18
   + 9

20 6
   + 6

Timed Tests: Addition and Subtraction © 2004 Creative Teaching Press

Subtraction Record Sheet

Test	Time	Score	Test	Time	Score
Warm-Up –2			Warm-Up –12		
Practice –2			Practice –12		
Challenge –2			Challenge –12		
Warm-Up –3			Warm-Up –13		
Practice –3			Practice –13		
Challenge –3			Challenge –13		
Warm-Up –4			Warm-Up –14		
Practice –4			Practice –14		
Challenge –4			Challenge –14		
Warm-Up –5			Warm-Up –15		
Practice –5			Practice –15		
Challenge –5			Challenge –15		
Warm-Up –6			Warm-Up –16		
Practice –6			Practice –16		
Challenge –6			Challenge –16		
Warm-Up –7			Warm-Up –17		
Practice –7			Practice –17		
Challenge –7			Challenge –17		
Warm-Up –8			Warm-Up –18		
Practice –8			Practice –18		
Challenge –8			Challenge –18		
Warm-Up –9			Warm-Up –19		
Practice –9			Practice –19		
Challenge –9			Challenge –19		
Warm-Up –10			Review 1–20A		
Practice –10			Review 1–20B		
Challenge –10			Addition and Subtraction 1–20A		
Warm-Up –11			Addition and Subtraction 1–20B		
Practice –11			Addition and Subtraction 1–20C		
Challenge –11			Addition and Subtraction 1–20D		

Name _____   Date _____

Goal Time _____   Actual Time _____   Score _____

 # **Subtraction Warm-Up**   –2

① 7
  – 2
—

② 15
  – 2
—

③ 6
  – 2
—

④ 10
  – 2
—

⑤ 17
  – 2
—

⑥ 18
  – 2
—

⑦ 22
  – 2
—

⑧ 14
  – 2
—

⑨ 5
  – 2
—

⑩ 13
  – 2
—

⑪ 9
  – 2
—

⑫ 20
  – 2
—

⑬ 3
  – 2
—

⑭ 11
  – 2
—

⑮ 16
  – 2
—

⑯ 4
  – 2
—

⑰ 12
  – 2
—

⑱ 19
  – 2
—

⑲ 8
  – 2
—

⑳ 21
  – 2
—

Timed Tests: Addition and Subtraction © 2004 Creative Teaching Press

Goal Time _____ Actual Time _____ Score _____

Subtraction Practice   −2

1 $\begin{array}{r} 3 \\ -\ 2 \\ \hline \end{array}$ **2** $\begin{array}{r} 9 \\ -\ 2 \\ \hline \end{array}$ **3** $\begin{array}{r} 20 \\ -\ 2 \\ \hline \end{array}$ **4** $\begin{array}{r} 19 \\ -\ 2 \\ \hline \end{array}$ **5** $\begin{array}{r} 14 \\ -\ 2 \\ \hline \end{array}$

6 $\begin{array}{r} 11 \\ -\ 2 \\ \hline \end{array}$ **7** $\begin{array}{r} 4 \\ -\ 2 \\ \hline \end{array}$ **8** $\begin{array}{r} 15 \\ -\ 2 \\ \hline \end{array}$ **9** $\begin{array}{r} 12 \\ -\ 2 \\ \hline \end{array}$ **10** $\begin{array}{r} 17 \\ -\ 2 \\ \hline \end{array}$

11 $\begin{array}{r} 21 \\ -\ 2 \\ \hline \end{array}$ **12** $\begin{array}{r} 7 \\ -\ 2 \\ \hline \end{array}$ **13** $\begin{array}{r} 5 \\ -\ 2 \\ \hline \end{array}$ **14** $\begin{array}{r} 18 \\ -\ 2 \\ \hline \end{array}$ **15** $\begin{array}{r} 10 \\ -\ 2 \\ \hline \end{array}$

16 $\begin{array}{r} 8 \\ -\ 2 \\ \hline \end{array}$ **17** $\begin{array}{r} 13 \\ -\ 2 \\ \hline \end{array}$ **18** $\begin{array}{r} 6 \\ -\ 2 \\ \hline \end{array}$ **19** $\begin{array}{r} 22 \\ -\ 2 \\ \hline \end{array}$ **20** $\begin{array}{r} 16 \\ -\ 2 \\ \hline \end{array}$

Name _____ Date _____

Goal Time _____ Actual Time _____ Score _____

 # Subtraction Challenge −2

1 ☐
$$-\ 2$$
$$\overline{7}$$

2 ☐
$$-\ 2$$
$$\overline{12}$$

3 ☐
$$-\ 2$$
$$\overline{3}$$

4 ☐
$$-\ 2$$
$$\overline{15}$$

5 ☐
$$-\ 2$$
$$\overline{10}$$

6 ☐
$$-\ 2$$
$$\overline{14}$$

7 ☐
$$-\ 2$$
$$\overline{1}$$

8 ☐
$$-\ 2$$
$$\overline{19}$$

9 ☐
$$-\ 2$$
$$\overline{8}$$

10 ☐
$$-\ 2$$
$$\overline{20}$$

11 ☐
$$-\ 2$$
$$\overline{5}$$

12 ☐
$$-\ 2$$
$$\overline{17}$$

13 ☐
$$-\ 2$$
$$\overline{11}$$

14 ☐
$$-\ 2$$
$$\overline{4}$$

15 ☐
$$-\ 2$$
$$\overline{6}$$

16 ☐
$$-\ 2$$
$$\overline{9}$$

17 ☐
$$-\ 2$$
$$\overline{16}$$

18 ☐
$$-\ 2$$
$$\overline{2}$$

19 ☐
$$-\ 2$$
$$\overline{18}$$

20 ☐
$$-\ 2$$
$$\overline{13}$$

Timed Tests: Addition and Subtraction © 2004 Creative Teaching Press

Name _____ Date _____

Goal Time _____ Actual Time _____ Score _____

 # Subtraction Warm-Up −3

1 $\begin{array}{r} 5 \\ -\ 3 \\ \hline \end{array}$ **2** $\begin{array}{r} 14 \\ -\ 3 \\ \hline \end{array}$ **3** $\begin{array}{r} 20 \\ -\ 3 \\ \hline \end{array}$ **4** $\begin{array}{r} 7 \\ -\ 3 \\ \hline \end{array}$ **5** $\begin{array}{r} 11 \\ -\ 3 \\ \hline \end{array}$

6 $\begin{array}{r} 17 \\ -\ 3 \\ \hline \end{array}$ **7** $\begin{array}{r} 10 \\ -\ 3 \\ \hline \end{array}$ **8** $\begin{array}{r} 23 \\ -\ 3 \\ \hline \end{array}$ **9** $\begin{array}{r} 21 \\ -\ 3 \\ \hline \end{array}$ **10** $\begin{array}{r} 18 \\ -\ 3 \\ \hline \end{array}$

11 $\begin{array}{r} 12 \\ -\ 3 \\ \hline \end{array}$ **12** $\begin{array}{r} 6 \\ -\ 3 \\ \hline \end{array}$ **13** $\begin{array}{r} 19 \\ -\ 3 \\ \hline \end{array}$ **14** $\begin{array}{r} 15 \\ -\ 3 \\ \hline \end{array}$ **15** $\begin{array}{r} 8 \\ -\ 3 \\ \hline \end{array}$

16 $\begin{array}{r} 9 \\ -\ 3 \\ \hline \end{array}$ **17** $\begin{array}{r} 16 \\ -\ 3 \\ \hline \end{array}$ **18** $\begin{array}{r} 22 \\ -\ 3 \\ \hline \end{array}$ **19** $\begin{array}{r} 4 \\ -\ 3 \\ \hline \end{array}$ **20** $\begin{array}{r} 13 \\ -\ 3 \\ \hline \end{array}$

Name _____ Date _____

Subtraction Practice   −3

1 23
    − 3

2 12
    − 3

3 18
    − 3

4   7
    − 3

5 15
    − 3

6 22
    − 3

7 16
    − 3

8   4
    − 3

9 19
    − 3

10 10
    − 3

11   6
    − 3

12 20
    − 3

13   8
    − 3

14 13
    − 3

15   9
    − 3

16 14
    − 3

17 11
    − 3

18 17
    − 3

19 21
    − 3

20   5
    − 3

Timed Tests: Addition and Subtraction © 2004 Creative Teaching Press

Name _____   Date _____

Goal Time _____   Actual Time _____   Score _____

 # Subtraction Challenge   −3

❶ □
−3
‾‾
1

❷ □
−3
‾‾
13

❸ □
−3
‾‾
6

❹ □
−3
‾‾
10

❺ □
−3
‾‾
17

❻ □
−3
‾‾
8

❼ □
−3
‾‾
15

❽ □
−3
‾‾
20

❾ □
−3
‾‾
5

❿ □
−3
‾‾
18

⓫ □
−3
‾‾
11

⓬ □
−3
‾‾
2

⓭ □
−3
‾‾
19

⓮ □
−3
‾‾
7

⓯ □
−3
‾‾
14

⓰ □
−3
‾‾
16

⓱ □
−3
‾‾
4

⓲ □
−3
‾‾
9

⓳ □
−3
‾‾
12

⓴ □
−3
‾‾
3

Name _____   Date _____

Goal Time _____   Actual Time _____   Score _____

 # Subtraction Warm-Up   −4

1   13
     − 4

2   15
     − 4

3   7
     − 4

4   19
     − 4

5   23
     − 4

6   6
     − 4

7   12
     − 4

8   22
     − 4

9   8
     − 4

10   11
      − 4

11   9
      − 4

12   20
      − 4

13   24
      − 4

14   17
      − 4

15   21
      − 4

16   18
      − 4

17   16
      − 4

18   10
      − 4

19   14
      − 4

20   5
      − 4

Name _____   Date _____

Goal Time _____   Actual Time _____   Score _____

 # Subtraction Practice   −4

1 7
  − 4

2 10
  − 4

3 19
  − 4

4 13
  − 4

5 22
  − 4

6 15
  − 4

7 20
  − 4

8 5
  − 4

9 11
  − 4

10 18
  − 4

11 6
  − 4

12 14
  − 4

13 17
  − 4

14 9
  − 4

15 23
  − 4

16 12
  − 4

17 8
  − 4

18 21
  − 4

19 24
  − 4

20 16
  − 4

Name _____ Date _____

Goal Time _____ Actual Time _____ Score _____

 # Subtraction Challenge −4

❶
$$\begin{array}{r} \square \\ -\ 4 \\ \hline 13 \end{array}$$

❷
$$\begin{array}{r} \square \\ -\ 4 \\ \hline 7 \end{array}$$

❸
$$\begin{array}{r} \square \\ -\ 4 \\ \hline 1 \end{array}$$

❹
$$\begin{array}{r} \square \\ -\ 4 \\ \hline 15 \end{array}$$

❺
$$\begin{array}{r} \square \\ -\ 4 \\ \hline 4 \end{array}$$

❻
$$\begin{array}{r} \square \\ -\ 4 \\ \hline 2 \end{array}$$

❼
$$\begin{array}{r} \square \\ -\ 4 \\ \hline 9 \end{array}$$

❽
$$\begin{array}{r} \square \\ -\ 4 \\ \hline 18 \end{array}$$

❾
$$\begin{array}{r} \square \\ -\ 4 \\ \hline 6 \end{array}$$

❿
$$\begin{array}{r} \square \\ -\ 4 \\ \hline 19 \end{array}$$

⓫
$$\begin{array}{r} \square \\ -\ 4 \\ \hline 16 \end{array}$$

⓬
$$\begin{array}{r} \square \\ -\ 4 \\ \hline 5 \end{array}$$

⓭
$$\begin{array}{r} \square \\ -\ 4 \\ \hline 12 \end{array}$$

⓮
$$\begin{array}{r} \square \\ -\ 4 \\ \hline 17 \end{array}$$

⓯
$$\begin{array}{r} \square \\ -\ 4 \\ \hline 10 \end{array}$$

⓰
$$\begin{array}{r} \square \\ -\ 4 \\ \hline 11 \end{array}$$

⓱
$$\begin{array}{r} \square \\ -\ 4 \\ \hline 20 \end{array}$$

⓲
$$\begin{array}{r} \square \\ -\ 4 \\ \hline 8 \end{array}$$

⓳
$$\begin{array}{r} \square \\ -\ 4 \\ \hline 3 \end{array}$$

⓴
$$\begin{array}{r} \square \\ -\ 4 \\ \hline 14 \end{array}$$

Timed Tests: Addition and Subtraction © 2004 Creative Teaching Press

Name _____   Date _____

Goal Time _____   Actual Time _____   Score _____

 Subtraction Warm-Up   −5

1   19
     − 5

2   12
     − 5

3   23
     − 5

4   8
     − 5

5   16
     − 5

6   10
     − 5

7   7
     − 5

8   24
     − 5

9   14
     − 5

10   20
      − 5

11   17
      − 5

12   13
      − 5

13   21
      − 5

14   25
      − 5

15   11
      − 5

16   6
      − 5

17   22
      − 5

18   15
      − 5

19   18
      − 5

20   9
      − 5

Name _____ Date _____

Goal Time _____ Actual Time _____ Score _____

 # Subtraction Practice −5

1 15
− 5

2 10
− 5

3 23
− 5

4 12
− 5

5 8
− 5

6 19
− 5

7 13
− 5

8 6
− 5

9 20
− 5

10 17
− 5

11 7
− 5

12 16
− 5

13 22
− 5

14 11
− 5

15 24
− 5

16 18
− 5

17 21
− 5

18 9
− 5

19 14
− 5

20 25
− 5

Timed Tests: Addition and Subtraction © 2004 Creative Teaching Press

Name _____ Date _____

Goal Time _____ Actual Time _____ Score _____

 Subtraction Challenge   **−5**

1 ☐
− 5
13

2 ☐
− 5
8

3 ☐
− 5
3

4 ☐
− 5
12

5 ☐
− 5
17

6 ☐
− 5
10

7 ☐
− 5
6

8 ☐
− 5
14

9 ☐
− 5
7

10 ☐
− 5
20

11 ☐
− 5
5

12 ☐
− 5
2

13 ☐
− 5
9

14 ☐
− 5
19

15 ☐
− 5
11

16 ☐
− 5
16

17 ☐
− 5
15

18 ☐
− 5
1

19 ☐
− 5
4

20 ☐
− 5
18

Name _____ Date _____

Goal Time _____ Actual Time _____ Score _____

 # Subtraction Warm-Up −6

1 9
   − 6

2 16
   − 6

3 25
   − 6

4 13
   − 6

5 21
   − 6

6 18
   − 6

7 11
   − 6

8 20
   − 6

9 17
   − 6

10 26
    − 6

11 14
    − 6

12 7
    − 6

13 22
    − 6

14 10
    − 6

15 24
    − 6

16 23
    − 6

17 12
    − 6

18 19
    − 6

19 8
    − 6

20 15
    − 6

Timed Tests: Addition and Subtraction © 2004 Creative Teaching Press

Subtraction Practice −6

1 17
   − 6

2 12
   − 6

3 20
   − 6

4 10
   − 6

5 22
   − 6

6 14
   − 6

7   9
   − 6

8 25
   − 6

9 15
   − 6

10   7
   − 6

11 21
   − 6

12 26
   − 6

13 24
   − 6

14 13
   − 6

15 18
   − 6

16 19
   − 6

17 11
   − 6

18 16
   − 6

19   8
   − 6

20 23
   − 6

Name _____ Date _____

Goal Time _____ Actual Time _____ Score _____

 # Subtraction Challenge −6

1 ☐
  − 6
 ____
  15

2 ☐
  − 6
 ____
   4

3 ☐
  − 6
 ____
  17

4 ☐
  − 6
 ____
   9

5 ☐
  − 6
 ____
  19

6 ☐
  − 6
 ____
   8

7 ☐
  − 6
 ____
  11

8 ☐
  − 6
 ____
   5

9 ☐
  − 6
 ____
   2

10 ☐
  − 6
 ____
  14

11 ☐
  − 6
 ____
  13

12 ☐
  − 6
 ____
   1

13 ☐
  − 6
 ____
  20

14 ☐
  − 6
 ____
   6

15 ☐
  − 6
 ____
  12

16 ☐
  − 6
 ____
   7

17 ☐
  − 6
 ____
  16

18 ☐
  − 6
 ____
   3

19 ☐
  − 6
 ____
  10

20 ☐
  − 6
 ____
  18

Timed Tests: Addition and Subtraction © 2004 Creative Teaching Press

Name _____ Date _____

Goal Time _____ Actual Time _____ Score _____

 # Subtraction Warm-Up   –7

1 16
     – 7

2 22
     – 7

3 10
     – 7

4 25
     – 7

5 14
     – 7

6   8
     – 7

7 15
     – 7

8 19
     – 7

9 11
     – 7

10 24
      – 7

11 12
      – 7

12 23
      – 7

13 27
      – 7

14 18
      – 7

15 20
      – 7

16 21
      – 7

17 17
      – 7

18 26
      – 7

19 13
      – 7

20   9
      – 7

Timed Tests: Addition and Subtraction © 2004 Creative Teaching Press

Name _____ Date _____

Goal Time _____ Actual Time _____ Score _____

 # Subtraction Practice   −7

1 27
− 7

2 20
− 7

3 25
− 7

4 12
− 7

5 23
− 7

6 18
− 7

7 11
− 7

8 22
− 7

9 15
− 7

10 19
− 7

11 24
− 7

12 13
− 7

13 26
− 7

14 8
− 7

15 17
− 7

16 16
− 7

17 9
− 7

18 14
− 7

19 21
− 7

20 10
− 7

Timed Tests: Addition and Subtraction © 2004 Creative Teaching Press

Name _____ Date _____

Goal Time _____ Actual Time _____ Score _____

 # Subtraction Challenge −7

❶
$$\begin{array}{r} \boxed{} \\ -\ 7 \\ \hline 20 \end{array}$$

❷
$$\begin{array}{r} \boxed{} \\ -\ 7 \\ \hline 3 \end{array}$$

❸
$$\begin{array}{r} \boxed{} \\ -\ 7 \\ \hline 11 \end{array}$$

❹
$$\begin{array}{r} \boxed{} \\ -\ 7 \\ \hline 15 \end{array}$$

❺
$$\begin{array}{r} \boxed{} \\ -\ 7 \\ \hline 8 \end{array}$$

❻
$$\begin{array}{r} \boxed{} \\ -\ 7 \\ \hline 13 \end{array}$$

❼
$$\begin{array}{r} \boxed{} \\ -\ 7 \\ \hline 1 \end{array}$$

❽
$$\begin{array}{r} \boxed{} \\ -\ 7 \\ \hline 16 \end{array}$$

❾
$$\begin{array}{r} \boxed{} \\ -\ 7 \\ \hline 5 \end{array}$$

❿
$$\begin{array}{r} \boxed{} \\ -\ 7 \\ \hline 18 \end{array}$$

⓫
$$\begin{array}{r} \boxed{} \\ -\ 7 \\ \hline 6 \end{array}$$

⓬
$$\begin{array}{r} \boxed{} \\ -\ 7 \\ \hline 19 \end{array}$$

⓭
$$\begin{array}{r} \boxed{} \\ -\ 7 \\ \hline 9 \end{array}$$

⓮
$$\begin{array}{r} \boxed{} \\ -\ 7 \\ \hline 2 \end{array}$$

⓯
$$\begin{array}{r} \boxed{} \\ -\ 7 \\ \hline 14 \end{array}$$

⓰
$$\begin{array}{r} \boxed{} \\ -\ 7 \\ \hline 17 \end{array}$$

⓱
$$\begin{array}{r} \boxed{} \\ -\ 7 \\ \hline 4 \end{array}$$

⓲
$$\begin{array}{r} \boxed{} \\ -\ 7 \\ \hline 10 \end{array}$$

⓳
$$\begin{array}{r} \boxed{} \\ -\ 7 \\ \hline 7 \end{array}$$

⓴
$$\begin{array}{r} \boxed{} \\ -\ 7 \\ \hline 12 \end{array}$$

Name _____ Date _____

 # Subtraction Warm-Up   −8

1   18
     − 8

2   11
     − 8

3   26
     − 8

4   15
     − 8

5   22
     − 8

6   13
     − 8

7   16
     − 8

8   24
     − 8

9   10
     − 8

10   19
      − 8

11   20
      − 8

12   9
      − 8

13   27
      − 8

14   21
      − 8

15   14
      − 8

16   23
      − 8

17   12
      − 8

18   25
      − 8

19   28
      − 8

20   17
      − 8

Name _____ Date _____

Goal Time _____ Actual Time _____ Score _____

 # Subtraction Practice −8

1 28
   − 8

2 21
   − 8

3 25
   − 8

4 19
   − 8

5 12
   − 8

6 17
   − 8

7 10
   − 8

8 27
   − 8

9 15
   − 8

10 24
   − 8

11 14
   − 8

12 23
   − 8

13 18
   − 8

14 11
   − 8

15 22
   − 8

16 26
   − 8

17 13
   − 8

18 20
   − 8

19 9
   − 8

20 16
   − 8

Name _____ Date _____

Goal Time _____ Actual Time _____ Score _____

 # Subtraction Challenge   −8

1 □
− 8
‾‾‾‾
16

2 □
− 8
‾‾‾‾
12

3 □
− 8
‾‾‾‾
4

4 □
− 8
‾‾‾‾
20

5 □
− 8
‾‾‾‾
8

6 □
− 8
‾‾‾‾
2

7 □
− 8
‾‾‾‾
10

8 □
− 8
‾‾‾‾
17

9 □
− 8
‾‾‾‾
6

10 □
− 8
‾‾‾‾
14

11 □
− 8
‾‾‾‾
7

12 □
− 8
‾‾‾‾
13

13 □
− 8
‾‾‾‾
1

14 □
− 8
‾‾‾‾
15

15 □
− 8
‾‾‾‾
18

16 □
− 8
‾‾‾‾
19

17 □
− 8
‾‾‾‾
5

18 □
− 8
‾‾‾‾
11

19 □
− 8
‾‾‾‾
9

20 □
− 8
‾‾‾‾
3

Subtraction Warm-Up   −9

1   11
      − 9

2   23
      − 9

3   15
      − 9

4   26
      − 9

5   19
      − 9

6   28
      − 9

7   17
      − 9

8   27
      − 9

9   12
      − 9

10   25
       − 9

11   20
       − 9

12   14
       − 9

13   29
       − 9

14   24
       − 9

15   16
       − 9

16   22
       − 9

17   13
       − 9

18   21
       − 9

19   18
       − 9

20   10
       − 9

Name _____ Date _____

Goal Time _____ Actual Time _____ Score _____

 # Subtraction Practice −9

1 15
   − 9
   ___

2 26
   − 9
   ___

3 19
   − 9
   ___

4 27
   − 9
   ___

5 22
   − 9
   ___

6 23
   − 9
   ___

7 16
   − 9
   ___

8 11
   − 9
   ___

9 25
   − 9
   ___

10 28
   − 9
   ___

11 12
   − 9
   ___

12 20
   − 9
   ___

13 17
   − 9
   ___

14 29
   − 9
   ___

15 14
   − 9
   ___

16 24
   − 9
   ___

17 13
   − 9
   ___

18 21
   − 9
   ___

19 10
   − 9
   ___

20 18
   − 9
   ___

Timed Tests: Addition and Subtraction © 2004 Creative Teaching Press

Name _____ Date _____

Goal Time _____ Actual Time _____ Score _____

 # Subtraction Challenge −9

1 ☐
$$\begin{array}{r} -\ 9 \\ \hline 16 \end{array}$$

2 ☐
$$\begin{array}{r} -\ 9 \\ \hline 7 \end{array}$$

3 ☐
$$\begin{array}{r} -\ 9 \\ \hline 13 \end{array}$$

4 ☐
$$\begin{array}{r} -\ 9 \\ \hline 5 \end{array}$$

5 ☐
$$\begin{array}{r} -\ 9 \\ \hline 10 \end{array}$$

6 ☐
$$\begin{array}{r} -\ 9 \\ \hline 3 \end{array}$$

7 ☐
$$\begin{array}{r} -\ 9 \\ \hline 11 \end{array}$$

8 ☐
$$\begin{array}{r} -\ 9 \\ \hline 18 \end{array}$$

9 ☐
$$\begin{array}{r} -\ 9 \\ \hline 2 \end{array}$$

10 ☐
$$\begin{array}{r} -\ 9 \\ \hline 19 \end{array}$$

11 ☐
$$\begin{array}{r} -\ 9 \\ \hline 6 \end{array}$$

12 ☐
$$\begin{array}{r} -\ 9 \\ \hline 15 \end{array}$$

13 ☐
$$\begin{array}{r} -\ 9 \\ \hline 1 \end{array}$$

14 ☐
$$\begin{array}{r} -\ 9 \\ \hline 12 \end{array}$$

15 ☐
$$\begin{array}{r} -\ 9 \\ \hline 8 \end{array}$$

16 ☐
$$\begin{array}{r} -\ 9 \\ \hline 14 \end{array}$$

17 ☐
$$\begin{array}{r} -\ 9 \\ \hline 9 \end{array}$$

18 ☐
$$\begin{array}{r} -\ 9 \\ \hline 20 \end{array}$$

19 ☐
$$\begin{array}{r} -\ 9 \\ \hline 4 \end{array}$$

20 ☐
$$\begin{array}{r} -\ 9 \\ \hline 17 \end{array}$$

Name _____ Date _____

Goal Time _____ Actual Time _____ Score _____

 # Subtraction Warm-Up −10

1
$$\begin{array}{r} 25 \\ -10 \\ \hline \end{array}$$

2
$$\begin{array}{r} 19 \\ -10 \\ \hline \end{array}$$

3
$$\begin{array}{r} 30 \\ -10 \\ \hline \end{array}$$

4
$$\begin{array}{r} 15 \\ -10 \\ \hline \end{array}$$

5
$$\begin{array}{r} 22 \\ -10 \\ \hline \end{array}$$

6
$$\begin{array}{r} 27 \\ -10 \\ \hline \end{array}$$

7
$$\begin{array}{r} 23 \\ -10 \\ \hline \end{array}$$

8
$$\begin{array}{r} 14 \\ -10 \\ \hline \end{array}$$

9
$$\begin{array}{r} 20 \\ -10 \\ \hline \end{array}$$

10
$$\begin{array}{r} 12 \\ -10 \\ \hline \end{array}$$

11
$$\begin{array}{r} 29 \\ -10 \\ \hline \end{array}$$

12
$$\begin{array}{r} 18 \\ -10 \\ \hline \end{array}$$

13
$$\begin{array}{r} 11 \\ -10 \\ \hline \end{array}$$

14
$$\begin{array}{r} 24 \\ -10 \\ \hline \end{array}$$

15
$$\begin{array}{r} 26 \\ -10 \\ \hline \end{array}$$

16
$$\begin{array}{r} 16 \\ -10 \\ \hline \end{array}$$

17
$$\begin{array}{r} 21 \\ -10 \\ \hline \end{array}$$

18
$$\begin{array}{r} 17 \\ -10 \\ \hline \end{array}$$

19
$$\begin{array}{r} 28 \\ -10 \\ \hline \end{array}$$

20
$$\begin{array}{r} 13 \\ -10 \\ \hline \end{array}$$

Timed Tests: Addition and Subtraction © 2004 Creative Teaching Press

Subtraction Practice −10

1 21
−10

2 14
−10

3 20
−10

4 17
−10

5 22
−10

6 13
−10

7 25
−10

8 29
−10

9 15
−10

10 26
−10

11 18
−10

12 27
−10

13 12
−10

14 24
−10

15 19
−10

16 23
−10

17 16
−10

18 30
−10

19 28
−10

20 11
−10

Name _____ Date _____

Goal Time _____ Actual Time _____ Score _____

 Subtraction Challenge −10

❶ ☐
−10
12

❷ ☐
−10
9

❸ ☐
−10
17

❹ ☐
−10
5

❺ ☐
−10
19

❻ ☐
−10
4

❼ ☐
−10
16

❽ ☐
−10
7

❾ ☐
−10
2

❿ ☐
−10
11

⓫ ☐
−10
18

⓬ ☐
−10
1

⓭ ☐
−10
10

⓮ ☐
−10
8

⓯ ☐
−10
14

⓰ ☐
−10
15

⓱ ☐
−10
6

⓲ ☐
−10
13

⓳ ☐
−10
20

⓴ ☐
−10
3

Name _____ Date _____

Goal Time _____ Actual Time _____ Score _____

 # Subtraction Warm-Up −11

1 23
     −11

2 18
     −11

3 24
     −11

4 30
     −11

5 16
     −11

6 27
     −11

7 15
     −11

8 12
     −11

9 21
     −11

10 28
      −11

11 17
      −11

12 26
      −11

13 20
      −11

14 29
      −11

15 14
      −11

16 13
      −11

17 22
      −11

18 31
      −11

19 19
      −11

20 25
      −11

Name _____ Date _____

Goal Time _____ Actual Time _____ Score _____

 # Subtraction Practice −11

1
30
−11

2
23
−11

3
16
−11

4
25
−11

5
19
−11

6
12
−11

7
18
−11

8
31
−11

9
15
−11

10
26
−11

11
27
−11

12
21
−11

13
20
−11

14
24
−11

15
13
−11

16
29
−11

17
28
−11

18
17
−11

19
14
−11

20
22
−11

Timed Tests: Addition and Subtraction © 2004 Creative Teaching Press

Subtraction Challenge   −11

1 ☐
−11
1

2 ☐
−11
15

3 ☐
−11
10

4 ☐
−11
4

5 ☐
−11
13

6 ☐
−11
17

7 ☐
−11
3

8 ☐
−11
12

9 ☐
−11
7

10 ☐
−11
19

11 ☐
−11
5

12 ☐
−11
16

13 ☐
−11
20

14 ☐
−11
8

15 ☐
−11
2

16 ☐
−11
14

17 ☐
−11
11

18 ☐
−11
6

19 ☐
−11
18

20 ☐
−11
9

Name _____ Date _____

Goal Time _____ Actual Time _____ Score _____

 # Subtraction Warm-Up   −12

1   16
       −12

2   30
       −12

3   28
       −12

4   23
       −12

5   15
       −12

6   22
       −12

7   14
       −12

8   32
       −12

9   17
       −12

10   25
        −12

11   19
        −12

12   26
        −12

13   21
        −12

14   13
        −12

15   29
        −12

16   24
        −12

17   31
        −12

18   18
        −12

19   20
        −12

20   27
        −12

Timed Tests: Addition and Subtraction © 2004 Creative Teaching Press

Name _____ Date _____

Goal Time _____ Actual Time _____ Score _____

 # Subtraction Practice   −12

1 32
−12

2 18
−12

3 24
−12

4 16
−12

5 27
−12

6 15
−12

7 21
−12

8 28
−12

9 17
−12

10 22
−12

11 30
−12

12 14
−12

13 29
−12

14 23
−12

15 20
−12

16 19
−12

17 25
−12

18 26
−12

19 13
−12

20 31
−12

Name _____ Date _____

Goal Time _____ Actual Time _____ Score _____

 # Subtraction Challenge −12

1 ☐
−12
9

2 ☐
−12
14

3 ☐
−12
3

4 ☐
−12
11

5 ☐
−12
19

6 ☐
−12
5

7 ☐
−12
16

8 ☐
−12
7

9 ☐
−12
13

10 ☐
−12
1

11 ☐
−12
18

12 ☐
−12
2

13 ☐
−12
10

14 ☐
−12
6

15 ☐
−12
17

16 ☐
−12
12

17 ☐
−12
8

18 ☐
−12
20

19 ☐
−12
4

20 ☐
−12
15

Name _____ Date _____

Goal Time _____ Actual Time _____ Score _____

 Subtraction Warm-Up −13

1 21
−13

2 14
−13

3 30
−13

4 25
−13

5 18
−13

6 17
−13

7 29
−13

8 22
−13

9 32
−13

10 27
−13

11 24
−13

12 33
−13

13 15
−13

14 31
−13

15 20
−13

16 26
−13

17 19
−13

18 23
−13

19 28
−13

20 16
−13

Name _____ Date _____

Goal Time _____ Actual Time _____ Score _____

 # Subtraction Practice   −13

① 28
−13

② 30
−13

③ 25
−13

④ 31
−13

⑤ 33
−13

⑥ 16
−13

⑦ 26
−13

⑧ 19
−13

⑨ 29
−13

⑩ 32
−13

⑪ 27
−13

⑫ 24
−13

⑬ 15
−13

⑭ 20
−13

⑮ 21
−13

⑯ 14
−13

⑰ 18
−13

⑱ 23
−13

⑲ 17
−13

⑳ 22
−13

Timed Tests: Addition and Subtraction © 2004 Creative Teaching Press

Name _____ Date _____

Goal Time _____ Actual Time _____ Score _____

 Subtraction Challenge   −13

❶ ☐
−13
5

❷ ☐
−13
17

❸ ☐
−13
9

❹ ☐
−13
3

❺ ☐
−13
11

❻ ☐
−13
14

❼ ☐
−13
19

❽ ☐
−13
1

❾ ☐
−13
18

❿ ☐
−13
7

⓫ ☐
−13
8

⓬ ☐
−13
16

⓭ ☐
−13
4

⓮ ☐
−13
13

⓯ ☐
−13
12

⓰ ☐
−13
2

⓱ ☐
−13
20

⓲ ☐
−13
10

⓳ ☐
−13
6

⓴ ☐
−13
15

Name _____ Date _____

Goal Time _____ Actual Time _____ Score _____

 # Subtraction Warm-Up −14

1 17
−14

2 23
−14

3 29
−14

4 20
−14

5 26
−14

6 25
−14

7 19
−14

8 33
−14

9 15
−14

10 31
−14

11 28
−14

12 34
−14

13 22
−14

14 30
−14

15 18
−14

16 21
−14

17 32
−14

18 16
−14

19 27
−14

20 24
−14

Timed Tests: Addition and Subtraction © 2004 Creative Teaching Press

Name _____ Date _____

Goal Time _____ Actual Time _____ Score _____

Subtraction Practice   −14

① 30
     −14

② 27
     −14

③ 29
     −14

④ 20
     −14

⑤ 26
     −14

⑥ 23
     −14

⑦ 18
     −14

⑧ 16
     −14

⑨ 15
     −14

⑩ 17
     −14

⑪ 19
     −14

⑫ 28
     −14

⑬ 34
     −14

⑭ 22
     −14

⑮ 32
     −14

⑯ 25
     −14

⑰ 21
     −14

⑱ 24
     −14

⑲ 33
     −14

⑳ 31
     −14

Name _____ Date _____

Goal Time _____ Actual Time _____ Score _____

 # Subtraction Challenge −14

① ☐
−14
2

② ☐
−14
9

③ ☐
−14
15

④ ☐
−14
19

⑤ ☐
−14
6

⑥ ☐
−14
13

⑦ ☐
−14
5

⑧ ☐
−14
11

⑨ ☐
−14
18

⑩ ☐
−14
3

⑪ ☐
−14
7

⑫ ☐
−14
14

⑬ ☐
−14
1

⑭ ☐
−14
10

⑮ ☐
−14
20

⑯ ☐
−14
4

⑰ ☐
−14
17

⑱ ☐
−14
12

⑲ ☐
−14
8

⑳ ☐
−14
16

Timed Tests: Addition and Subtraction © 2004 Creative Teaching Press

Goal Time _____   Actual Time _____   Score _____

Subtraction Warm-Up   −15

1 26
    −15

2 21
    −15

3 29
    −15

4 18
    −15

5 24
    −15

6 17
    −15

7 32
    −15

8 20
    −15

9 35
    −15

10 28
     −15

11 33
     −15

12 19
     −15

13 27
     −15

14 23
     −15

15 30
     −15

16 22
     −15

17 31
     −15

18 25
     −15

19 34
     −15

20 16
     −15

Name _____ Date _____

Goal Time _____ Actual Time _____ Score _____

 # Subtraction Practice   −15

1   23
      −15

2   20
      −15

3   30
      −15

4   33
      −15

5   22
      −15

6   31
      −15

7   17
      −15

8   32
      −15

9   19
      −15

10 28
      −15

11 26
      −15

12 24
      −15

13 29
      −15

14 34
      −15

15 27
      −15

16 21
      −15

17 18
      −15

18 35
      −15

19 25
      −15

20 16
      −15

Timed Tests: Addition and Subtraction © 2004 Creative Teaching Press

Name _____ Date _____

Goal Time _____ Actual Time _____ Score _____

 # Subtraction Challenge −15

1 ☐
−15
‾‾1‾

2 ☐
−15
‾‾8‾

3 ☐
−15
‾12‾

4 ☐
−15
‾‾6‾

5 ☐
−15
‾18‾

6 ☐
−15
‾‾5‾

7 ☐
−15
‾10‾

8 ☐
−15
‾16‾

9 ☐
−15
‾‾3‾

10 ☐
−15
‾14‾

11 ☐
−15
‾13‾

12 ☐
−15
‾‾2‾

13 ☐
−15
‾20‾

14 ☐
−15
‾‾9‾

15 ☐
−15
‾17‾

16 ☐
−15
‾19‾

17 ☐
−15
‾‾7‾

18 ☐
−15
‾15‾

19 ☐
−15
‾‾4‾

20 ☐
−15
‾11‾

Name _____ Date _____

Goal Time _____ Actual Time _____ Score _____

 # Subtraction Warm-Up −16

1 20
  −16

2 28
  −16

3 36
  −16

4 22
  −16

5 34
  −16

6 24
  −16

7 17
  −16

8 32
  −16

9 30
  −16

10 26
   −16

11 33
   −16

12 21
   −16

13 29
   −16

14 18
   −16

15 35
   −16

16 27
   −16

17 19
   −16

18 23
   −16

19 31
   −16

20 25
   −16

Timed Tests: Addition and Subtraction © 2004 Creative Teaching Press

Goal Time _____ Actual Time _____ Score _____

 # Subtraction Practice   −16

1 21
  −16

2 28
  −16

3 32
  −16

4 26
  −16

5 18
  −16

6 17
  −16

7 30
  −16

8 23
  −16

9 34
  −16

10 36
  −16

11 25
  −16

12 35
  −16

13 19
  −16

14 33
  −16

15 24
  −16

16 22
  −16

17 27
  −16

18 31
  −16

19 20
  −16

20 29
  −16

Name _____ Date _____

Goal Time _____ Actual Time _____ Score _____

 # Subtraction Challenge −16

① ☐
−16
4

② ☐
−16
13

③ ☐
−16
7

④ ☐
−16
15

⑤ ☐
−16
1

⑥ ☐
−16
9

⑦ ☐
−16
17

⑧ ☐
−16
5

⑨ ☐
−16
11

⑩ ☐
−16
20

⑪ ☐
−16
14

⑫ ☐
−16
2

⑬ ☐
−16
8

⑭ ☐
−16
16

⑮ ☐
−16
19

⑯ ☐
−16
18

⑰ ☐
−16
6

⑱ ☐
−16
10

⑲ ☐
−16
3

⑳ ☐
−16
12

Timed Tests: Addition and Subtraction © 2004 Creative Teaching Press

Subtraction Warm-Up   −17

1 29
      −17

2 18
      −17

3 35
      −17

4 23
      −17

5 27
      −17

6 25
      −17

7 33
      −17

8 21
      −17

9 37
      −17

10 31
       −17

11 22
       −17

12 34
       −17

13 30
       −17

14 19
       −17

15 28
       −17

16 20
       −17

17 26
       −17

18 32
       −17

19 36
       −17

20 24
       −17

Subtraction Practice   −17

1 22
     −17

2 30
     −17

3 27
     −17

4 34
     −17

5 20
     −17

6 19
     −17

7 33
     −17

8 24
     −17

9 37
     −17

10 31
      −17

11 36
      −17

12 23
      −17

13 29
      −17

14 18
      −17

15 26
      −17

16 25
      −17

17 32
      −17

18 21
      −17

19 28
      −17

20 35
      −17

Timed Tests: Addition and Subtraction © 2004 Creative Teaching Press

Subtraction Challenge -17

❶ ☐
-17
$\overline{7}$

❷ ☐
-17
$\overline{13}$

❸ ☐
-17
$\overline{5}$

❹ ☐
-17
$\overline{16}$

❺ ☐
-17
$\overline{10}$

❻ ☐
-17
$\overline{4}$

❼ ☐
-17
$\overline{8}$

❽ ☐
-17
$\overline{17}$

❾ ☐
-17
$\overline{3}$

❿ ☐
-17
$\overline{14}$

⓫ ☐
-17
$\overline{20}$

⓬ ☐
-17
$\overline{11}$

⓭ ☐
-17
$\overline{2}$

⓮ ☐
-17
$\overline{15}$

⓯ ☐
-17
$\overline{19}$

⓰ ☐
-17
$\overline{1}$

⓱ ☐
-17
$\overline{9}$

⓲ ☐
-17
$\overline{18}$

⓳ ☐
-17
$\overline{6}$

⓴ ☐
-17
$\overline{12}$

Name _____ Date _____

Goal Time _____ Actual Time _____ Score _____

 # Subtraction Warm-Up −18

1 22
−18

2 30
−18

3 26
−18

4 34
−18

5 27
−18

6 33
−18

7 25
−18

8 38
−18

9 21
−18

10 29
−18

11 37
−18

12 19
−18

13 36
−18

14 31
−18

15 24
−18

16 23
−18

17 28
−18

18 35
−18

19 20
−18

20 32
−18

Timed Tests: Addition and Subtraction © 2004 Creative Teaching Press

Goal Time _____ Actual Time _____ Score _____

 Subtraction Practice   −18

1 19
  −18

2 37
  −18

3 22
  −18

4 28
  −18

5 31
  −18

6 30
  −18

7 26
  −18

8 36
  −18

9 24
  −18

10 34
   −18

11 32
   −18

12 21
   −18

13 38
   −18

14 29
   −18

15 25
   −18

16 23
   −18

17 35
   −18

18 27
   −18

19 33
   −18

20 20
   −18

Name _____   Date _____

Goal Time _____   Actual Time _____   Score _____

 Subtraction Challenge   −18

❶ ☐
−18
11

❷ ☐
−18
4

❸ ☐
−18
19

❹ ☐
−18
7

❺ ☐
−18
13

❻ ☐
−18
5

❼ ☐
−18
8

❽ ☐
−18
14

❾ ☐
−18
17

❿ ☐
−18
2

⓫ ☐
−18
16

⓬ ☐
−18
12

⓭ ☐
−18
3

⓮ ☐
−18
20

⓯ ☐
−18
9

⓰ ☐
−18
18

⓱ ☐
−18
1

⓲ ☐
−18
10

⓳ ☐
−18
15

⓴ ☐
−18
6

Timed Tests: Addition and Subtraction © 2004 Creative Teaching Press

Subtraction Warm-Up   –19

1   23
      –19

2   30
      –19

3   37
      –19

4   24
      –19

5   35
      –19

6   27
      –19

7   39
      –19

8   20
      –19

9   33
      –19

10 26
      –19

11 21
      –19

12 34
      –19

13 28
      –19

14 31
      –19

15 38
      –19

16 32
      –19

17 25
      –19

18 36
      –19

19 22
      –19

20 29
      –19

Name _____ Date _____

Goal Time _____ Actual Time _____ Score _____

 # Subtraction Practice   −19

1 32
−19

2 24
−19

3 35
−19

4 20
−19

5 39
−19

6 28
−19

7 37
−19

8 22
−19

9 33
−19

10 25
−19

11 21
−19

12 30
−19

13 36
−19

14 29
−19

15 34
−19

16 27
−19

17 31
−19

18 38
−19

19 23
−19

20 26
−19

Name _____ Date _____

Goal Time _____ Actual Time _____ Score _____

 # Subtraction Challenge −19

1 ☐
−19
9

2 ☐
−19
17

3 ☐
−19
12

4 ☐
−19
6

5 ☐
−19
20

6 ☐
−19
11

7 ☐
−19
4

8 ☐
−19
16

9 ☐
−19
2

10 ☐
−19
14

11 ☐
−19
8

12 ☐
−19
15

13 ☐
−19
1

14 ☐
−19
10

15 ☐
−19
7

16 ☐
−19
3

17 ☐
−19
19

18 ☐
−19
13

19 ☐
−19
5

20 ☐
−19
18

Goal Time _____ Actual Time _____ Score _____

 # Subtraction Review   1–20A

1 10
    – 7

2 19
    –12

3 14
    – 9

4   9
    – 4

5   7
    – 3

6 17
    –15

7 25
    –11

8   3
    – 2

9 20
    –16

10 18
    – 5

11 23
    –11

12 12
    – 6

13 26
    –19

14 21
    –13

15 25
    –14

16 13
    – 8

17 22
    –10

18   8
    – 7

19 29
    –17

20 30
    –18

Timed Tests: Addition and Subtraction © 2004 Creative Teaching Press

Name _____   Date _____

Goal Time _____   Actual Time _____   Score _____

 Subtraction Review   1–20B

1   15
  −12

2   24
  −19

3   19
  −18

4   6
  − 3

5   12
  − 8

6   14
  − 7

7   20
  −13

8   22
  −17

9   4
  − 2

10   17
   − 9

11   11
   − 6

12   9
   − 4

13   13
   − 5

14   29
   −16

15   16
   −10

16   27
   −14

17   23
   −11

18   25
   −15

19   7
   − 2

20   14
   − 5

Name _____ Date _____

Goal Time _____ Actual Time _____ Score _____

 # Addition and Subtraction
Review   1–20A

1 $\begin{array}{r} 5 \\ +\ 9 \\ \hline \end{array}$ **2** $\begin{array}{r} 31 \\ -13 \\ \hline \end{array}$ **3** $\begin{array}{r} 11 \\ +\ 3 \\ \hline \end{array}$ **4** $\begin{array}{r} 9 \\ -\ 6 \\ \hline \end{array}$ **5** $\begin{array}{r} 13 \\ -\ 8 \\ \hline \end{array}$

6 $\begin{array}{r} 16 \\ -10 \\ \hline \end{array}$ **7** $\begin{array}{r} 8 \\ +\ 8 \\ \hline \end{array}$ **8** $\begin{array}{r} 13 \\ +\ 7 \\ \hline \end{array}$ **9** $\begin{array}{r} 24 \\ -14 \\ \hline \end{array}$ **10** $\begin{array}{r} 18 \\ +19 \\ \hline \end{array}$

11 $\begin{array}{r} 7 \\ -\ 5 \\ \hline \end{array}$ **12** $\begin{array}{r} 12 \\ +\ 2 \\ \hline \end{array}$ **13** $\begin{array}{r} 18 \\ -\ 7 \\ \hline \end{array}$ **14** $\begin{array}{r} 15 \\ +17 \\ \hline \end{array}$ **15** $\begin{array}{r} 22 \\ -16 \\ \hline \end{array}$

16 $\begin{array}{r} 26 \\ -18 \\ \hline \end{array}$ **17** $\begin{array}{r} 14 \\ +\ 4 \\ \hline \end{array}$ **18** $\begin{array}{r} 17 \\ +\ 5 \\ \hline \end{array}$ **19** $\begin{array}{r} 10 \\ -\ 4 \\ \hline \end{array}$ **20** $\begin{array}{r} 10 \\ +\ 6 \\ \hline \end{array}$

Timed Tests: Addition and Subtraction © 2004 Creative Teaching Press

Name _____ Date _____

Goal Time _____ Actual Time _____ Score _____

 # Addition and Subtraction
Review   1–20B

1
$$\begin{array}{r} 5 \\ +\ 5 \\ \hline \end{array}$$

2
$$\begin{array}{r} 23 \\ -11 \\ \hline \end{array}$$

3
$$\begin{array}{r} 7 \\ +\ 9 \\ \hline \end{array}$$

4
$$\begin{array}{r} 6 \\ +18 \\ \hline \end{array}$$

5
$$\begin{array}{r} 9 \\ -\ 2 \\ \hline \end{array}$$

6
$$\begin{array}{r} 28 \\ -15 \\ \hline \end{array}$$

7
$$\begin{array}{r} 13 \\ +15 \\ \hline \end{array}$$

8
$$\begin{array}{r} 12 \\ -\ 9 \\ \hline \end{array}$$

9
$$\begin{array}{r} 4 \\ +\ 8 \\ \hline \end{array}$$

10
$$\begin{array}{r} 25 \\ -16 \\ \hline \end{array}$$

11
$$\begin{array}{r} 14 \\ -\ 6 \\ \hline \end{array}$$

12
$$\begin{array}{r} 19 \\ +19 \\ \hline \end{array}$$

13
$$\begin{array}{r} 26 \\ -13 \\ \hline \end{array}$$

14
$$\begin{array}{r} 16 \\ +\ 9 \\ \hline \end{array}$$

15
$$\begin{array}{r} 2 \\ +16 \\ \hline \end{array}$$

16
$$\begin{array}{r} 32 \\ -19 \\ \hline \end{array}$$

17
$$\begin{array}{r} 11 \\ +11 \\ \hline \end{array}$$

18
$$\begin{array}{r} 13 \\ -\ 7 \\ \hline \end{array}$$

19
$$\begin{array}{r} 10 \\ +14 \\ \hline \end{array}$$

20
$$\begin{array}{r} 36 \\ -14 \\ \hline \end{array}$$

Addition and Subtraction Review   1–20C

1 22
    −12

2   4
    +17

3 15
    − 9

4 20
    − 8

5 12
    +13

6 19
    + 2

7 24
    −10

8 10
    +10

9 26
    −14

10 14
    +19

11   9
    − 5

12 20
    +16

13 35
    −18

14   5
    + 7

15 19
    −13

16 17
    +17

17 11
    − 7

18 15
    + 8

19 18
    + 3

20 30
    −16

Timed Tests: Addition and Subtraction © 2004 Creative Teaching Press

Addition and Subtraction
Review   1–20D

1 11
 +13

2   8
  – 4

3 12
 +20

4 30
 –17

5   4
  +19

6 28
 –15

7 18
 +18

8 15
  – 9

9   3
  + 5

10 24
 –12

11   7
  + 8

12 10
  – 6

13   6
 +14

14 27
 –11

15 10
  + 9

16 36
 –19

17 12
 +12

18 11
  – 8

19 16
 +17

20 32
 –18

Answer Key

Warm-Up +2
1. 10	2. 5	3. 8	4. 6	5. 16
6. 12	7. 3	8. 4	9. 21	10. 7
11. 15	12. 11	13. 17	14. 13	15. 19
16. 14	17. 20	18. 9	19. 22	20. 18

Warm-Up +5
1. 6	2. 20	3. 16	4. 12	5. 23
6. 22	7. 8	8. 24	9. 18	10. 13
11. 19	12. 14	13. 10	14. 21	15. 7
16. 9	17. 25	18. 17	19. 11	20. 15

Warm-Up +8
1. 15	2. 20	3. 17	4. 24	5. 9
6. 27	7. 10	8. 22	9. 14	10. 25
11. 12	12. 18	13. 26	14. 11	15. 23
16. 21	17. 13	18. 28	19. 16	20. 19

Practice +2
1. 7	2. 12	3. 10	4. 21	5. 17
6. 5	7. 3	8. 16	9. 14	10. 20
11. 8	12. 19	13. 11	14. 4	15. 13
16. 18	17. 6	18. 22	19. 15	20. 9

Practice +5
1. 19	2. 7	3. 25	4. 13	5. 21
6. 23	7. 9	8. 18	9. 11	10. 15
11. 16	12. 12	13. 20	14. 6	15. 17
16. 24	17. 8	18. 14	19. 22	20. 10

Practice +8
1. 10	2. 24	3. 16	4. 12	5. 27
6. 15	7. 19	8. 13	9. 22	10. 26
11. 9	12. 17	13. 28	14. 20	15. 14
16. 21	17. 25	18. 11	19. 23	20. 18

Challenge +2
1. 10	2. 3	3. 16	4. 9	5. 14
6. 6	7. 17	8. 1	9. 12	10. 19
11. 7	12. 5	13. 11	14. 8	15. 2
16. 15	17. 18	18. 4	19. 13	20. 20

Challenge +5
1. 5	2. 16	3. 3	4. 19	5. 10
6. 13	7. 18	8. 7	9. 20	10. 1
11. 11	12. 2	13. 14	14. 17	15. 8
16. 4	17. 9	18. 15	19. 6	20. 12

Challenge +8
1. 8	2. 15	3. 2	4. 18	5. 7
6. 10	7. 6	8. 19	9. 4	10. 11
11. 13	12. 1	13. 16	14. 9	15. 20
16. 17	17. 5	18. 12	19. 3	20. 14

Warm-Up +3
1. 6	2. 16	3. 9	4. 4	5. 17
6. 13	7. 10	8. 7	9. 22	10. 12
11. 8	12. 11	13. 19	14. 14	15. 20
16. 15	17. 23	18. 21	19. 5	20. 18

Warm-Up +6
1. 15	2. 24	3. 8	4. 20	5. 14
6. 26	7. 7	8. 22	9. 11	10. 18
11. 10	12. 19	13. 13	14. 25	15. 16
16. 23	17. 12	18. 17	19. 21	20. 9

Warm-Up +9
1. 19	2. 22	3. 13	4. 27	5. 15
6. 16	7. 25	8. 10	9. 18	10. 23
11. 11	12. 28	13. 20	14. 14	15. 26
16. 17	17. 24	18. 12	19. 29	20. 21

Practice +3
1. 12	2. 13	3. 6	4. 22	5. 19
6. 17	7. 18	8. 16	9. 15	10. 21
11. 11	12. 8	13. 20	14. 4	15. 9
16. 14	17. 5	18. 23	19. 10	20. 7

Practice +6
1. 20	2. 24	3. 9	4. 16	5. 12
6. 7	7. 18	8. 11	9. 14	10. 23
11. 22	12. 25	13. 8	14. 13	15. 17
16. 10	17. 21	18. 26	19. 15	20. 19

Practice +9
1. 29	2. 17	3. 25	4. 12	5. 28
6. 21	7. 23	8. 14	9. 19	10. 26
11. 27	12. 11	13. 18	14. 15	15. 22
16. 13	17. 20	18. 16	19. 24	20. 10

Challenge +3
1. 15	2. 7	3. 19	4. 4	5. 11
6. 9	7. 16	8. 1	9. 13	10. 18
11. 3	12. 10	13. 20	14. 6	15. 14
16. 12	17. 5	18. 8	19. 17	20. 2

Challenge +6
1. 1	2. 12	3. 6	4. 20	5. 9
6. 7	7. 15	8. 11	9. 4	10. 17
11. 18	12. 3	13. 16	14. 10	15. 2
16. 5	17. 13	18. 19	19. 8	20. 14

Challenge +9
1. 5	2. 13	3. 11	4. 2	5. 18
6. 7	7. 15	8. 3	9. 20	10. 9
11. 16	12. 1	13. 12	14. 6	15. 14
16. 10	17. 8	18. 17	19. 4	20. 19

Warm-Up +4
1. 19	2. 7	3. 16	4. 23	5. 10
6. 6	7. 21	8. 14	9. 12	10. 18
11. 17	12. 5	13. 9	14. 20	15. 13
16. 24	17. 11	18. 15	19. 22	20. 8

Warm-Up +7
1. 9	2. 17	3. 26	4. 12	5. 20
6. 21	7. 13	8. 8	9. 24	10. 15
11. 14	12. 25	13. 16	14. 18	15. 10
16. 23	17. 11	18. 19	19. 27	20. 22

Warm-Up +10
1. 13	2. 27	3. 18	4. 30	5. 23
6. 16	7. 14	8. 28	9. 22	10. 17
11. 24	12. 20	13. 12	14. 15	15. 29
16. 11	17. 21	18. 19	19. 26	20. 25

Practice +4
1. 18	2. 6	3. 23	4. 12	5. 13
6. 8	7. 22	8. 15	9. 9	10. 19
11. 14	12. 10	13. 21	14. 16	15. 5
16. 17	17. 24	18. 20	19. 7	20. 11

Practice +7
1. 16	2. 20	3. 12	4. 23	5. 18
6. 13	7. 24	8. 15	9. 9	10. 27
11. 25	12. 10	13. 17	14. 22	15. 11
16. 8	17. 21	18. 14	19. 26	20. 19

Practice +10
1. 19	2. 17	3. 22	4. 28	5. 13
6. 14	7. 29	8. 11	9. 25	10. 23
11. 24	12. 12	13. 21	14. 18	15. 26
16. 15	17. 20	18. 16	19. 27	20. 30

Challenge +4
1. 3	2. 17	3. 11	4. 7	5. 15
6. 12	7. 9	8. 20	9. 5	10. 18
11. 6	12. 14	13. 1	14. 16	15. 10
16. 13	17. 4	18. 19	19. 8	20. 2

Challenge +7
1. 15	2. 6	3. 18	4. 11	5. 3
6. 9	7. 1	8. 13	9. 5	10. 19
11. 17	12. 10	13. 7	14. 2	15. 14
16. 4	17. 20	18. 12	19. 16	20. 8

Challenge +10
1. 11	2. 7	3. 18	4. 5	5. 13
6. 4	7. 19	8. 2	9. 15	10. 1
11. 9	12. 14	13. 8	14. 20	15. 16
16. 12	17. 6	18. 17	19. 3	20. 10

Warm-Up +11

1. 17	2. 31	3. 15	4. 28	5. 25
6. 30	7. 19	8. 12	9. 23	10. 21
11. 13	12. 26	13. 20	14. 16	15. 24
16. 22	17. 14	18. 29	19. 18	20. 27

Practice +11

1. 16	2. 21	3. 19	4. 30	5. 24
6. 27	7. 14	8. 25	9. 17	10. 22
11. 15	12. 20	13. 23	14. 12	15. 31
16. 29	17. 13	18. 26	19. 18	20. 28

Challenge +11

1. 11	2. 19	3. 1	4. 14	5. 8
6. 5	7. 15	8. 7	9. 20	10. 4
11. 18	12. 3	13. 10	14. 16	15. 12
16. 9	17. 6	18. 13	19. 2	20. 17

Warm-Up +12

1. 15	2. 23	3. 29	4. 18	5. 26
6. 31	7. 19	8. 14	9. 27	10. 32
11. 17	12. 24	13. 21	14. 30	15. 22
16. 13	17. 20	18. 28	19. 25	20. 16

Practice +12

1. 30	2. 20	3. 25	4. 16	5. 32
6. 17	7. 24	8. 22	9. 28	10. 13
11. 26	12. 14	13. 29	14. 18	15. 31
16. 19	17. 23	18. 15	19. 27	20. 21

Challenge +12

1. 12	2. 19	3. 1	4. 15	5. 5
6. 2	7. 11	8. 6	9. 17	10. 8
11. 10	12. 16	13. 20	14. 13	15. 3
16. 14	17. 9	18. 4	19. 18	20. 7

Warm-Up +13

1. 19	2. 32	3. 18	4. 24	5. 29
6. 26	7. 17	8. 33	9. 22	10. 15
11. 21	12. 28	13. 14	14. 27	15. 20
16. 16	17. 23	18. 30	19. 25	20. 31

Practice +13

1. 24	2. 16	3. 29	4. 33	5. 19
6. 18	7. 21	8. 30	9. 14	10. 25
11. 23	12. 17	13. 32	14. 20	15. 22
16. 27	17. 15	18. 31	19. 28	20. 26

Challenge +13

1. 1	2. 9	3. 17	4. 5	5. 13
6. 15	7. 4	8. 19	9. 8	10. 11
11. 12	12. 7	13. 2	14. 14	15. 18
16. 3	17. 16	18. 10	19. 20	20. 6

Warm-Up +14

1. 26	2. 20	3. 33	4. 16	5. 29
6. 18	7. 34	8. 21	9. 32	10. 24
11. 31	12. 22	13. 25	14. 15	15. 27
16. 23	17. 17	18. 28	19. 19	20. 30

Practice +14

1. 23	2. 25	3. 22	4. 17	5. 32
6. 15	7. 30	8. 19	9. 26	10. 33
11. 21	12. 27	13. 16	14. 20	15. 24
16. 18	17. 28	18. 29	19. 34	20. 31

Challenge +14

1. 16	2. 3	3. 14	4. 6	5. 11
6. 5	7. 9	8. 17	9. 1	10. 19
11. 13	12. 4	13. 10	14. 18	15. 7
16. 12	17. 20	18. 2	19. 8	20. 15

Warm-Up +15

1. 20	2. 31	3. 18	4. 23	5. 25
6. 28	7. 21	8. 33	9. 35	10. 29
11. 26	12. 17	13. 30	14. 24	15. 32
16. 16	17. 34	18. 22	19. 19	20. 27

Practice +15

1. 24	2. 18	3. 29	4. 16	5. 23
6. 27	7. 22	8. 31	9. 20	10. 34
11. 21	12. 25	13. 17	14. 28	15. 32
16. 30	17. 19	18. 33	19. 26	20. 35

Challenge +15

1. 11	2. 7	3. 16	4. 3	5. 14
6. 2	7. 17	8. 5	9. 20	10. 9
11. 6	12. 15	13. 10	14. 1	15. 12
16. 13	17. 4	18. 19	19. 8	20. 18

Warm-Up +16

1. 17	2. 30	3. 23	4. 34	5. 21
6. 28	7. 25	8. 20	9. 31	10. 36
11. 22	12. 33	13. 24	14. 26	15. 18
16. 19	17. 32	18. 27	19. 29	20. 35

Practice +16

1. 25	2. 19	3. 32	4. 34	5. 21
6. 24	7. 22	8. 28	9. 20	10. 26
11. 30	12. 17	13. 33	14. 23	15. 29
16. 27	17. 36	18. 31	19. 18	20. 35

Challenge +16

1. 14	2. 16	3. 3	4. 19	5. 10
6. 4	7. 11	8. 20	9. 7	10. 15
11. 8	12. 2	13. 18	14. 5	15. 12
16. 13	17. 6	18. 17	19. 9	20. 1

Warm-Up +17

1. 22	2. 32	3. 27	4. 25	5. 30
6. 33	7. 21	8. 36	9. 19	10. 37
11. 26	12. 18	13. 35	14. 23	15. 28
16. 20	17. 29	18. 24	19. 31	20. 34

Practice +17

1. 19	2. 30	3. 37	4. 21	5. 34
6. 23	7. 26	8. 18	9. 36	10. 31
11. 29	12. 24	13. 20	14. 32	15. 27
16. 33	17. 22	18. 28	19. 25	20. 35

Challenge +17

1. 3	2. 9	3. 17	4. 6	5. 14
6. 11	7. 5	8. 18	9. 13	10. 1
11. 15	12. 2	13. 20	14. 8	15. 19
16. 7	17. 10	18. 16	19. 4	20. 12

Warm-Up +18

1. 27	2. 23	3. 30	4. 19	5. 34
6. 32	7. 20	8. 38	9. 25	10. 36
11. 22	12. 35	13. 26	14. 31	15. 28
16. 37	17. 24	18. 29	19. 21	20. 33

Practice +18

1. 24	2. 33	3. 20	4. 31	5. 37
6. 25	7. 32	8. 35	9. 21	10. 30
11. 38	12. 29	13. 34	14. 23	15. 36
16. 27	17. 19	18. 28	19. 22	20. 26

Challenge +18

1. 14	2. 3	3. 18	4. 9	5. 1
6. 5	7. 12	8. 7	9. 20	10. 15
11. 17	12. 2	13. 19	14. 6	15. 11
16. 10	17. 16	18. 4	19. 13	20. 8

Warm-Up +19

1. 24	2. 33	3. 26	4. 29	5. 21
6. 20	7. 39	8. 32	9. 28	10. 35
11. 31	12. 22	13. 37	14. 25	15. 30
16. 27	17. 34	18. 38	19. 23	20. 36

Practice +19

1. 22	2. 30	3. 27	4. 33	5. 37
6. 25	7. 34	8. 23	9. 38	10. 29
11. 20	12. 31	13. 39	14. 24	15. 35
16. 32	17. 28	18. 36	19. 21	20. 26

Challenge +19

1. 12	2. 1	3. 16	4. 5	5. 19
6. 3	7. 11	8. 9	9. 14	10. 7
11. 8	12. 15	13. 2	14. 17	15. 13
16. 18	17. 6	18. 20	19. 10	20. 4

Addition Review +1–10

1. 10 2. 11 3. 6 4. 14 5. 12
6. 12 7. 20 8. 7 9. 8 10. 8
11. 9 12. 12 13. 14 14. 6 15. 18
16. 11 17. 13 18. 17 19. 15 20. 10

Challenge –3

1. 4 2. 16 3. 9 4. 13 5. 20
6. 11 7. 18 8. 23 9. 8 10. 21
11. 14 12. 5 13. 22 14. 10 15. 17
16. 19 17. 7 18. 12 19. 15 20. 6

Challenge –6

1. 21 2. 10 3. 23 4. 15 5. 25
6. 14 7. 17 8. 11 9. 8 10. 20
11. 19 12. 7 13. 26 14. 12 15. 18
16. 13 17. 22 18. 9 19. 16 20. 24

Addition Review +11–20

1. 31 2. 30 3. 25 4. 33 5. 28
6. 31 7. 32 8. 36 9. 31 10. 34
11. 24 12. 31 13. 28 14. 35 15. 31
16. 37 17. 29 18. 29 19. 29 20. 35

Warm-Up –4

1. 9 2. 11 3. 3 4. 15 5. 19
6. 2 7. 8 8. 18 9. 4 10. 7
11. 5 12. 16 13. 20 14. 13 15. 17
16. 14 17. 12 18. 6 19. 10 20. 1

Warm-Up –7

1. 9 2. 15 3. 3 4. 18 5. 7
6. 1 7. 8 8. 12 9. 4 10. 17
11. 5 12. 16 13. 20 14. 11 15. 13
16. 14 17. 10 18. 19 19. 6 20. 2

Addition Review +1–20A

1. 15 2. 13 3. 26 4. 10 5. 18
6. 23 7. 19 8. 30 9. 20 10. 14
11. 16 12. 14 13. 16 14. 18 15. 16
16. 17 17. 22 18. 25 19. 17 20. 25

Practice –4

1. 3 2. 6 3. 15 4. 9 5. 18
6. 11 7. 16 8. 1 9. 7 10. 14
11. 2 12. 10 13. 13 14. 5 15. 19
16. 8 17. 4 18. 17 19. 20 20. 12

Practice –7

1. 20 2. 13 3. 18 4. 5 5. 16
6. 11 7. 4 8. 15 9. 8 10. 12
11. 17 12. 6 13. 19 14. 1 15. 10
16. 9 17. 2 18. 7 19. 14 20. 3

Addition Review +1–20B

1. 11 2. 15 3. 22 4. 17 5. 8
6. 23 7. 22 8. 14 9. 27 10. 23
11. 21 12. 18 13. 38 14. 13 15. 18
16. 39 17. 20 18. 28 19. 27 20. 12

Challenge –4

1. 17 2. 11 3. 5 4. 19 5. 8
6. 6 7. 13 8. 22 9. 10 10. 23
11. 20 12. 9 13. 16 14. 21 15. 14
16. 15 17. 24 18. 12 19. 7 20. 18

Challenge –7

1. 27 2. 10 3. 18 4. 22 5. 15
6. 20 7. 8 8. 23 9. 12 10. 25
11. 13 12. 26 13. 16 14. 9 15. 21
16. 24 17. 11 18. 17 19. 14 20. 19

Warm-Up –2

1. 5 2. 13 3. 4 4. 8 5. 15
6. 16 7. 20 8. 12 9. 3 10. 11
11. 7 12. 18 13. 1 14. 9 15. 14
16. 2 17. 10 18. 17 19. 6 20. 19

Warm-Up –5

1. 14 2. 7 3. 18 4. 3 5. 11
6. 5 7. 2 8. 19 9. 9 10. 15
11. 12 12. 8 13. 16 14. 20 15. 6
16. 1 17. 17 18. 10 19. 13 20. 4

Warm-Up –8

1. 10 2. 3 3. 18 4. 7 5. 14
6. 5 7. 8 8. 16 9. 2 10. 11
11. 12 12. 1 13. 19 14. 13 15. 6
16. 15 17. 4 18. 17 19. 20 20. 9

Practice –2

1. 1 2. 7 3. 18 4. 17 5. 12
6. 9 7. 2 8. 13 9. 10 10. 15
11. 19 12. 5 13. 3 14. 16 15. 8
16. 6 17. 11 18. 4 19. 20 20. 14

Practice –5

1. 10 2. 5 3. 18 4. 7 5. 3
6. 14 7. 8 8. 1 9. 15 10. 12
11. 2 12. 11 13. 17 14. 6 15. 19
16. 13 17. 16 18. 4 14. 9 20. 20

Practice –8

1. 20 2. 13 3. 17 4. 11 5. 4
6. 9 7. 2 8. 19 9. 7 10. 16
11. 6 12. 15 13. 10 14. 3 15. 14
16. 18 17. 5 18. 12 19. 1 20. 8

Challenge –2

1. 9 2. 14 3. 5 4. 17 5. 12
6. 16 7. 3 8. 21 9. 10 10. 22
11. 7 12. 19 13. 13 14. 6 15. 8
16. 11 17. 18 18. 4 19. 20 20. 15

Challenge –5

1. 18 2. 13 3. 8 4. 17 5. 22
6. 15 7. 11 8. 19 9. 12 10. 25
11. 10 12. 7 13. 14 14. 24 15. 16
16. 21 17. 20 18. 6 19. 9 20. 23

Challenge –8

1. 24 2. 20 3. 12 4. 28 5. 16
6. 10 7. 18 8. 25 9. 14 10. 22
11. 15 12. 21 13. 9 14. 23 15. 26
16. 27 17. 13 18. 19 19. 17 20. 11

Warm-Up –3

1. 2 2. 11 3. 17 4. 4 5. 8
6. 14 7. 7 8. 20 9. 18 10. 15
11. 9 12. 3 13. 16 14. 12 15. 5
16. 6 17. 13 18. 19 19. 1 20. 10

Warm-Up –6

1. 3 2. 10 3. 19 4. 7 5. 15
6. 12 7. 5 8. 14 9. 11 10. 20
11. 8 12. 1 13. 16 14. 4 15. 18
16. 17 17. 6 18. 13 19. 2 20. 9

Warm-Up –9

1. 2 2. 14 3. 6 4. 17 5. 10
6. 19 7. 8 8. 18 9. 3 10. 16
11. 11 12. 5 13. 20 14. 15 15. 7
16. 13 17. 4 18. 12 19. 9 15. 1

Practice –3

1. 20 2. 9 3. 15 4. 4 5. 12
6. 19 7. 13 8. 1 9. 16 10. 7
11. 3 12. 17 13. 5 14. 10 15. 6
16. 11 17. 8 18. 14 19. 18 20. 2

Practice –6

1. 11 2. 6 3. 14 4. 4 5. 16
6. 8 7. 3 8. 19 9. 9 10. 1
11. 15 12. 20 13. 18 14. 7 15. 12
16. 13 17. 5 18. 10 19. 2 20. 17

Practice –9

1. 6 2. 17 3. 10 4. 18 5. 13
6. 14 7. 7 8. 2 9. 16 10. 19
11. 3 12. 11 13. 8 14. 20 15. 5
16. 15 17. 4 18. 12 19. 1 20. 9

Challenge –9

1. 25	2. 16	3. 22	4. 14	5. 19
6. 12	7. 20	8. 27	9. 11	10. 28
11. 15	12. 24	13. 10	14. 21	15. 17
16. 23	17. 18	18. 29	19. 13	20. 26

Challenge –12

1. 21	2. 26	3. 15	4. 23	5. 31
6. 17	7. 28	8. 19	9. 25	10. 13
11. 30	12. 14	13. 22	14. 18	15. 29
16. 24	17. 20	18. 32	19. 16	20. 27

Challenge –15

1. 16	2. 23	3. 27	4. 21	5. 33
6. 20	7. 25	8. 31	9. 18	10. 29
11. 28	12. 17	13. 35	14. 24	15. 32
16. 34	17. 22	18. 30	19. 19	20. 26

Warm-Up –10

1. 15	2. 9	3. 20	4. 5	5. 12
6. 17	7. 13	8. 4	9. 10	10. 2
11. 19	12. 8	13. 1	14. 14	15. 16
16. 6	17. 11	18. 7	19. 18	20. 3

Warm-Up –13

1. 8	2. 1	3. 17	4. 12	5. 5
6. 4	7. 16	8. 9	9. 19	10. 14
11. 11	12. 20	13. 2	14. 18	15. 7
16. 13	17. 6	18. 10	19. 15	20. 3

Warm-Up –16

1. 4	2. 12	3. 20	4. 6	5. 18
6. 8	7. 1	8. 16	9. 14	10. 10
11. 17	12. 5	13. 13	14. 2	15. 19
16. 11	17. 3	18. 7	19. 15	20. 9

Practice –10

1. 11	2. 4	3. 10	4. 7	5. 12
6. 3	7. 15	8. 19	9. 5	10. 16
11. 8	12. 17	13. 2	14. 14	15. 9
16. 13	17. 6	18. 20	19. 18	20. 1

Practice –13

1. 15	2. 17	3. 12	4. 18	5. 20
6. 3	7. 13	8. 6	9. 16	10. 19
11. 14	12. 11	13. 2	14. 7	15. 8
16. 1	17. 5	18. 10	19. 4	20. 9

Practice –16

1. 5	2. 12	3. 16	4. 10	5. 2
6. 1	7. 14	8. 7	9. 18	10. 20
11. 9	12. 19	13. 3	14. 17	15. 8
16. 6	17. 11	18. 15	19. 4	20. 13

Challenge –10

1. 22	2. 19	3. 27	4. 15	5. 29
6. 14	7. 26	8. 17	9. 12	10. 21
11. 28	12. 11	13. 20	14. 18	15. 24
16. 25	17. 16	18. 23	19. 30	20. 13

Challenge –13

1. 18	2. 30	3. 22	4. 16	5. 24
6. 27	7. 32	8. 14	9. 31	10. 20
11. 21	12. 29	13. 17	14. 26	15. 25
16. 15	17. 33	18. 23	19. 19	20. 28

Challenge –16

1. 20	2. 29	3. 23	4. 31	5. 17
6. 25	7. 33	8. 21	9. 27	10. 36
11. 30	12. 18	13. 24	14. 32	15. 35
16. 34	17. 22	18. 26	19. 19	20. 28

Warm-Up –11

1. 12	2. 7	3. 13	4. 19	5. 5
6. 16	7. 4	8. 1	9. 10	10. 17
11. 6	12. 15	13. 9	14. 18	15. 3
16. 2	17. 11	18. 20	19. 8	20. 14

Warm-Up –14

1. 3	2. 9	3. 15	4. 6	5. 12
6. 11	7. 5	8. 19	9. 1	10. 17
11. 14	12. 20	13. 8	14. 16	15. 4
16. 7	17. 18	18. 2	19. 13	20. 10

Warm-Up –17

1. 12	2. 1	3. 18	4. 6	5. 10
6. 8	7. 16	8. 4	9. 20	10. 14
11. 5	12. 17	13. 13	14. 2	15. 11
16. 3	17. 9	18. 15	19. 19	20. 7

Practice –11

1. 19	2. 12	3. 5	4. 14	5. 8
6. 1	7. 7	8. 20	9. 4	10. 15
11. 16	12. 10	13. 9	14. 13	15. 2
16. 18	17. 17	18. 6	19. 3	20. 11

Practice –14

1. 16	2. 13	3. 15	4. 6	5. 12
6. 9	7. 4	8. 2	9. 1	10. 3
11. 5	12. 14	13. 20	14. 8	15. 18
16. 11	17. 7	18. 10	19. 19	20. 17

Practice –17

1. 5	2. 13	3. 10	4. 17	5. 3
6. 2	7. 16	8. 7	9. 20	10. 14
11. 19	12. 6	13. 12	14. 1	15. 9
16. 8	17. 15	18. 4	19. 11	20. 18

Challenge –11

1. 12	2. 26	3. 21	4. 15	5. 24
6. 28	7. 14	8. 23	9. 18	10. 30
11. 16	12. 27	13. 31	14. 19	15. 13
16. 25	17. 22	18. 17	19. 29	20. 20

Challenge –14

1. 16	2. 23	3. 29	4. 33	5. 20
6. 27	7. 19	8. 25	9. 32	10. 17
11. 21	12. 28	13. 15	14. 24	15. 34
16. 18	17. 31	18. 26	19. 22	20. 30

Challenge –17

1. 24	2. 30	3. 22	4. 33	5. 27
6. 21	7. 25	8. 34	9. 20	10. 31
11. 37	12. 28	13. 19	14. 32	15. 36
16. 18	17. 26	18. 35	19. 23	20. 29

Warm-Up –12

1. 4	2. 18	3. 16	4. 11	5. 3
6. 10	7. 2	8. 20	9. 5	10. 13
11. 7	12. 14	13. 9	14. 1	15. 17
16. 12	17. 19	18. 6	19. 8	20. 15

Warm-Up –15

1. 11	2. 6	3. 14	4. 3	5. 9
6. 2	7. 17	8. 5	9. 20	10. 13
11. 18	12. 4	13. 12	14. 8	15. 15
16. 7	17. 16	18. 10	19. 19	20. 1

Warm-Up –18

1. 4	2. 12	3. 8	4. 16	5. 9
6. 15	7. 7	8. 20	9. 3	10. 11
11. 19	12. 1	13. 18	14. 13	15. 6
16. 5	17. 10	18. 17	19. 2	20. 14

Practice –12

1. 20	2. 6	3. 12	4. 4	5. 15
6. 3	7. 9	8. 16	9. 5	10. 10
11. 18	12. 2	13. 17	14. 11	15. 8
16. 7	17. 13	18. 14	19. 1	20. 19

Practice –15

1. 8	2. 5	3. 15	4. 18	5. 7
6. 16	7. 2	8. 17	9. 4	10. 13
11. 11	12. 9	13. 14	14. 19	15. 12
16. 6	17. 3	18. 20	19. 10	20. 1

Practice –18

1. 1	2. 19	3. 4	4. 10	5. 13
6. 12	7. 8	8. 18	9. 6	10. 16
11. 14	12. 3	13. 20	14. 11	15. 7
16. 5	17. 17	18. 9	19. 15	20. 2

Challenge −18

1. 29	2. 22	3. 37	4. 25	5. 31
6. 23	7. 26	8. 32	9. 35	10. 20
11. 34	12. 30	13. 21	14. 38	15. 27
16. 36	17. 19	18. 28	19. 33	20. 24

Warm-Up −19

1. 4	2. 11	3. 18	4. 5	5. 16
6. 8	7. 20	8. 1	9. 14	10. 7
11. 2	12. 15	13. 9	14. 12	15. 19
16. 13	17. 6	18. 17	19. 3	20. 10

Practice −19

1. 13	2. 5	3. 16	4. 1	5. 20
6. 9	7. 18	8. 3	9. 14	10. 6
11. 2	12. 11	13. 17	14. 10	15. 15
16. 8	17. 12	18. 19	19. 4	20. 7

Challenge −19

1. 28	2. 36	3. 31	4. 25	5. 39
6. 30	7. 23	8. 35	9. 21	10. 33
11. 27	12. 34	13. 20	14. 29	15. 26
16. 22	17. 38	18. 32	19. 24	20. 37

Subtraction Review 1−20A

1. 3	2. 7	3. 5	4. 5	5. 4
6. 2	7. 14	8. 1	9. 4	10. 13
11. 12	12. 6	13. 7	14. 8	15. 11
16. 5	17. 12	18. 1	19. 12	20. 12

Subtraction Review 1−20B

1. 3	2. 5	3. 1	4. 3	5. 4
6. 7	7. 7	8. 5	9. 2	10. 8
11. 5	12. 5	13. 8	14. 13	15. 6
16. 13	17. 12	18. 10	19. 5	20. 9

Addition and Subtraction Review
1−20A

1. 14	2. 18	3. 14	4. 3	5. 5
6. 6	7. 16	8. 20	9. 10	10. 37
11. 2	12. 14	13. 11	14. 32	15. 6
16. 8	17. 18	18. 22	19. 6	20. 16

Addition and Subtraction Review
1−20B

1. 10	2. 12	3. 16	4. 24	5. 7
6. 13	7. 28	8. 3	9. 12	10. 9
11. 8	12. 38	13. 13	14. 25	15. 18
16. 13	17. 22	18. 6	19. 24	20. 22

Addition and Subtraction Review
1−20C

1. 10	2. 21	3. 6	4. 12	5. 25
6. 21	7. 14	8. 20	9. 12	10. 33
11. 4	12. 36	13. 17	14. 12	15. 6
16. 34	17. 4	18. 23	19. 21	20. 14

Addition and Subtraction Review
1−20D

1. 24	2. 4	3. 32	4. 13	5. 23
6. 13	7. 36	8. 6	9. 8	10. 12
11. 15	12. 4	13. 20	14. 16	15. 19
16. 17	17. 24	18. 3	19. 33	20. 14